Politics and Development of Contemporary China

Series Editors

Kevin G. Cai
University of Waterloo
Renison University College
Waterloo, ON, Canada

Pan Guang
Shanghai Center International Studies
Shanghai Academy of Social Sciences
Shanghai, China

Daniel C. Lynch
University of Southern California
School of International Relations
Los Angeles, CA, USA

As China's power grows, the search has begun in earnest for what superpower status will mean for the People's Republic of China as a nation as well as the impact of its new-found influence on the Asia-Pacific region and the global international order at large. By providing a venue for exciting and ground-breaking titles, the aim of this series is to explore the domestic and international implications of China's rise and transformation through a number of key areas including politics, development and foreign policy. The series will also give a strong voice to non-western perspectives on China's rise in order to provide a forum that connects and compares the views of academics from both the east and west reflecting the truly international nature of the discipline.

More information about this series at
http://www.palgrave.com/gp/series/14541

Liping Dai

Politics and Governance in Water Pollution Prevention in China

palgrave
macmillan

Liping Dai
Utrecht Centre for Water, Oceans
and Sustainability Law
Utrecht University
Utrecht, The Netherlands

Politics and Development of Contemporary China
ISBN 978-3-030-02873-2   ISBN 978-3-030-02874-9 (eBook)
https://doi.org/10.1007/978-3-030-02874-9

Library of Congress Control Number: 2018960249

This Palgrave Pivot imprint is published by the registered company Springer Nature Switzerland AG
The registered company address is: Gewerbestrasse 11, 6330 Cham, Switzerland

To my son, Dennis,
with love.

Acknowledgements

This book is based on my Ph.D. project of *China's Water Resource Law in Transition* conducted between 2011 and 2015 in Utrecht University, the Netherlands. I wish to express my special gratitude to Professor Marleen van Rijswick and Dr. Andrea Keessen, as my promoter and daily supervisor, respectively, they provide invaluable academic advice and personal encouragement throughout my four-year research at the Utrecht Centre for Water, Oceans and Sustainability Law.

It is my privilege to thank my husband, Adrian Butnaru, for his support, understanding and tolerance during the preparation of this book. And to my 9-month old son Dennis, this book would have not been finished without he going to bed early every day.

Many thanks also to the Publisher and its editorial staff for the valuable support in the realization of this book.

Contents

List of Tables

CHAPTER 1

Introduction

Abstract Along with the economic development, water pollution in China is developing from traditional pollution with conventional pollutants to a compound type of pollution with new and old pollutants interacting with each other. Traditional water management schemes are not able to meet these challenges. Improving water governance is a key topic in the political agenda in China. This book deploys a three-step diagnostic method—ten building blocks approach to explore China's current water governance for water pollution prevention and control. It consists of seven chapters, each discusses different building block(s) of water governance, including China's water quality knowledge, water policy discourses, public participation in water management, green path to water law, market-based approach and implementation of water goals.

Keywords Water pollution · China · Water governance

1 Background

Since 1978, China has been on the road towards a transition from a centrally planned economy to a market economy. The achievements are remarkable after almost three decades of effort. Industrialization in China has been unprecedented—no developing country has grown by more than 10% per year as China has, without interruption for two full

L. Dai, *Politics and Governance in Water Pollution Prevention in China*, Politics and Development of Contemporary China,
https://doi.org/10.1007/978-3-030-02874-9_1

decades (Eichengreen, 2013). The urbanization rate had reached 51% by 2011, and by 2030 nearly 60% of the population are expected to be living in urban areas (West, Schandl, Heyenga, & Chen, 2013). The living standards of ordinary citizens have improved significantly, so that millions of Chinese citizens have been lifted out of poverty.

However, pressure on the environment is likely to grow in response to the industrialization and urbanization. In the period between 2001 and 2005, about 54%, on average, of the seven main rivers in China contained water which was deemed to be unsafe for human consumption (Mundial, 2007), though this rate had dropped from 54% to around 9% by 2016 (Ministry of Environmental Protection, 2017). As a result of the economic development, water pollution in China is developing from traditional pollution with conventional pollutants to a compound type of pollution with new and old pollutants interacting with each other, and from industry-dominated pollution to the coexistence of industrial pollution and agricultural pollution (Wang, 2008). Traditional water management schemes are not able to meet these challenges.

Poor water quality not only threatens human and ecosystem health, increases treatment costs, and reduces the availability of safe water for drinking and other uses, but also limits economic productivity and development opportunities. The Chinese Government has paid great attention to the environmental issues in the past decade and made a number of efforts to address the challenges facing the water sector.

2 Water Governance

Water crisis is often a crisis of governance (Rogers & Hall, 2003). In its first World Water Development Report, the United Nations identified that "water crisis is essentially a crisis of governance and societies are facing a number of social, economic and political challenges on how to govern water more effectively" (World Water Assessment Programme, 2003). Notions of water governance (and of governance in general) have evolved over time. Early thinking about water governance was based on highly centralized systems that emphasized the role of governments in water management (Cooley et al., 2013). Today, the term is used more broadly; according to Rogers and Hall it refers to "the range of political, social, economic and administrative systems that are in place to develop and manage water resources, and the delivery of water services, at different levels of society" (Rogers & Hall, 2003). More specifically, water governance refers to formal and informal processes that allow for the determination and negotiation of objectives, setting of standards, and resolution of disputes among disparate voices in order to address challenges and meet objectives at local, sub-national, and national levels (Cooley et al., 2013).

Improving water governance is a key topic in the political agenda worldwide (OECD, 2011). Due to the complex nature of water systems (multilevel, multi-scale and multi-actor), a comprehensive water governance approach is needed in which different interests and uses of water are interconnected so that water policy and measurements are developed and implemented with the support of different stakeholder groups (van Rijswick, Edelenbos, Hellegers, Kok, & Kuks, 2014).

Governance can take many different forms depending on the economic, cultural and traditional political norms of a country and the behaviour of the legislature and legislators (Rogers & Hall, 2003). van Rijswick et al. apply a multidisciplinary perspective (a three-step diagnostic method and ten building blocks) to approach water governance (Fig. 1) (van Rijswick et al., 2014). Figure 1 illustrates the cyclic aspect of the method. Once the agreed service levels have been implemented, the loop might go back to the content as it can have consequences for the state of the water system, or when monitoring shows the need for revision of one of the building blocks (van Rijswick et al., 2014).

The "ten building block" assessment method assumes that water governance is sound when three main dimensions and the corresponding ten building blocks are dealt with. Sound water management requires knowledge of the water system in time and space and of values, principles and policy discourses. This knowledge is required for the organizational process to come to an agreed service level. The organizational process requires sufficient stakeholder involvement, insights into the trade-offs between social objectives, the attribution of responsibilities, authorization and the associated means as well as regulations and agreements. Finally, the agreed service level has to be implemented, which requires the engineering of infrastructure, monitoring, enforcement and conflict prevention and resolution (van Rijswick et al., 2014).

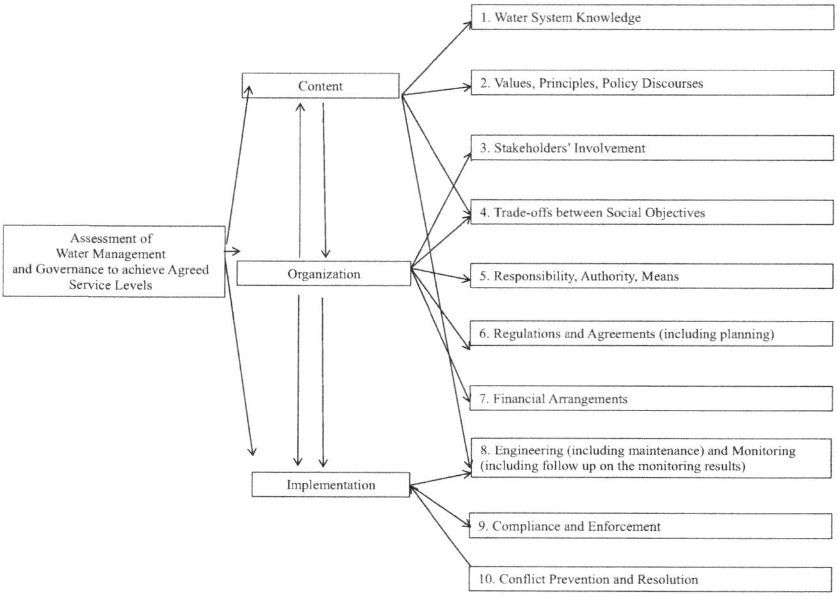

Fig. 1 A three-step diagnostic method—ten building blocks approach of water governance assessment

3 Structure of This Book

This book deploys the three-step diagnostic method—ten building blocks approach to explore China's current water governance for water pollution prevention and control. It is divided into seven chapters. Following this introduction, Chapter 2 discusses the current stage of water quality knowledge in China and provides detailed analysis of how the water problems are framed by the Chinese Government. More specifically, Chapter 2 discusses the blocks of "water system knowledge", "policy discourses" and "regulations".

Chapter 3 turns to the block of "stakeholder involvement". Good water governance and effective stakeholder involvement are inextricably linked since the collaborative processes can make or break planning and implementation of water management. This chapter focuses on public participation, a practice of stakeholder involvement, to explore whether the current regulations have offered the public opportunities to participate.

Chapter 4 discusses the transformation of water policy in recent China (the block of "trade-offs between social objectives"). As a result of the rapid industrialization and urbanization, water policy in China has been constantly changing during the past decades and a more sustainable development pathway is being developed. This chapter analyses how a sustainable, development-friendly approach to the utilization and management of water resources is reflected in China's water law.

For almost a decade China has been implementing "eco-compensation" mechanisms as an incentive approach to address water problems. Such an approach requires close cooperation between government and private sectors. Chapter 5 selects the Lake Tai watershed as a case study to show how this cooperation works at a regional level. It is concerned with the correlation of the blocks of "stakeholder involvement" and "financial arrangements".

China applies a different path of social development compared with some other countries. Chapter 6 aims to explain how this difference is reflected in the field of water governance. It explores how China applies political power as an institutional instrument to address its water quality challenges and implement its water goals. This chapter focuses on the block of "responsibility, authority, means".

The final chapter, Chapter 7, provides the summary and conclusion of this book.

Due to the author's profession, there is a certain cross-disciplinary barrier; thus, this book does not thoroughly explore each block or pay equal attention to each block and the analysis of each block, but it mainly studies the blocks from a regulatory perspective. However, it is also worth noting that the ten blocks are not independent from each other. A change to one component could affect other components. For example, if the policy discourses change, the formulation of the regulations or the allocation of responsibility will be affected and changed correspondingly. Therefore, readers should keep in mind that the relationship between the components in the ten blocks is dynamic, although they are categorized in different groups.

References

Cooley, H., Ajami, N., Ha, M.-L., Srinivasan, V., Morrison, J., Donnelly, K., & Christian-Smith, J. (2013). *Global Water Governance in the 21st Century*. Oakland, CA: Pacific Institute.

Eichengreen, B. (2013). *Chinese Industrialization and Its Discontents*. Retrieved from https://www.project-syndicate.org/commentary/barry-eichengreen-on-what-china-can-learn-from-nineteenth-century-britain?barrier=accesspaylog.

Ministry of Environmental Protection. (2017). *2016 China Environmental Situation Bulletin*. Retrieved from Beijing. http://www.mep.gov.cn/gkml/hbb/qt/201706/W020170605812243090317.pdf.

Mundial, B. (2007). *Cost of Pollution in China: Economic Estimates of Physical Damages*. Banco Mundial.

OECD. (2011). *Water Governance in OECD Countries: A Multi-Level Approach*. Retrieved from Online. http://www.oecd.org/cfe/regional-policy/watergovernanceinoecdcountriesamulti-levelapproach.htm.

Rogers, P., & Hall, A. W. (2003). *Effective Water Governance* (Vol. 7). Global Water Partnership Stockholm.

van Rijswick, H. F., Edelenbos, J., Hellegers, P., Kok, M., & Kuks, S. (2014). Ten Building Blocks for Sustainable Water Governance: An Integrated Method to Assess the Governance of Water. *Water International, 39*(5). https://doi.org/10.1080/02508060.2014.951828.

Wang, Y. (2008). China's Water Issues: Transition, Governance and Innovation. In J. Albiac & A. Dinar (Eds.), *The Management of Water Quality and Irrigation Technologies* (pp. 117–134). London: Earthscan.

West, J., Schandl, H., Heyenga, S., & Chen, S. (2013). *Resource Efficiency: Economics and Outlook for China*. United Nations Environment Programme.

World Water Assessment Programme. (2003). *Water for People, Water for Life: A Joint Report by the Twenty-Three UN Agencies Concerned with Freshwater*. UNESCO.

CHAPTER 2

Water Quality Knowledge and Water Policy Discourse in China

Abstract Good knowledge of current water status forms the basis for good governance for water quality. However, to understand the content of a water issue one needs to understand not only the facts such as water quality knowledge, but also the social and political context in which the water problems are constructed. This chapter aims to address these issues, it discusses the water quality knowledge at current stage in China, and analyses how water quality problem is framed by Chinese government through discussing its water-related policies and law in the last decade.

Keywords China · Water governance · Water quality · Water policies · Environmental law

1 Introduction

van Rijswick et al. believe that sound water governance requires good knowledge of the water system. In their research, "water system" is defined as the combination of natural physical resources (such as rivers, rainfall, seas, lakes) and man-made infrastructure (such as canals, pumping stations, reservoirs, flood defences). The system supports societal functions, such as domestic and industrial water use, irrigation, shipping, hydropower, safety, etc. (van Rijswick, Edelenbos, Hellegers, Kok, & Kuks, 2014). Water quality is one of the elements of the water system.

L. Dai, *Politics and Governance in Water Pollution Prevention in China*, Politics and Development of Contemporary China,
https://doi.org/10.1007/978-3-030-02874-9_2

In keeping with the scope of this book, this chapter focuses only on "water quality", not on the water system in a general sense.

Policy discourses can be considered as different ensembles of actors with specific storylines, frames, values and principles that emphasize certain aspects of water problems and policy measures (van Rijswick et al., 2014). Discourse provides the tools with which problems are construed (Hajer, 1993). Governance arrangements are in part constituted by and operate within and through discourses and discursive practices. Discourse analysis can help reveal "the how" of these processes and practices (Jeffares, Held, & Atkinson, 2010).

To understand the content of a water issue one needs to understand not only the facts (such as the water quality knowledge), but also the values, cultural, social and political background, and images that actors have developed in the course of time when dealing with water issues (van Rijswick et al., 2014). This chapter aims to address these issues: it first discusses the water quality knowledge at its current stage in China, and then analyses how the water quality problem is framed by the Chinese Government.

2 Water Quality Knowledge in China

Despite China being a water-rich country in terms of the total volume of water resources, the per capita availability, estimated in 2007, is only a quarter of the world average and amongst the lowest for a major country (Xie, 2008). Unfortunately, this limited water resource is compounded by high levels of pollution (Dai, 2015). Water pollution further exacerbates water scarcity. The overall cost of water scarcity associated with water pollution has amounted to 147 billion RMB ($21.3 billion) per year, or about 1% of GDP (Mundial, 2007). Pollution also poses serious risks to human health. Research indicates that approximately 190 million people fall ill, and 60,000 people die from a range of diseases and illnesses associated with water pollution in China each year (Mundial, 2007). After more than three decades of rapid economic growth, China's water pollution has become too serious to ignore.

In China, the former Ministry of Environmental Protection (which has been absorbed into the Ministry of Ecological Environment) is responsible for working out water environment monitoring standards, and for releasing the state's water environment situation by publishing a national bulletin on a yearly basis. According to the 2016 Bulletin, among the evaluated seven river basins, only 2.1% of surface water reached Class I. Classes II, III, IV and V accounted for 41.8, 27.3, 13.4, and 6.3%

respectively. Inferior Class V, which is unfit for potable use or human contact, had reached 9.1%. The inter-provincial waters with water quality classified as better than Class III and inferior Class V represented 67.1 and 17.1%, respectively, based on 544 important interprovincial sectors evaluated. The main parameters exceeding the standards in evaluated river basins were Chemical Oxygen Demand (COD), Total Phosphorus (TP) and 5-day Biochemical Oxygen Demand (BOD5), and in inter-provincial waters they were COD, TP, and Ammonia Nitrogen (Ministry of Environmental Protection, 2017). The contributors of COD discharges, from the greatest to the least, were agricultural wastewater, domestic sewage and industrial wastewater; the contributors of Ammonia Nitrogen discharges in turn were domestic sewage, agricultural wastewater and industrial wastewater (Ministry of Environmental Protection, 2017).

3 Policy Discourse for Water Quality Governance in China

As China is a one-party state, politics and law in China deeply influence and penetrate each other. The Communist Party has a profound influence on social development, as well as on water quality management. This makes it necessary to discuss both Party policy and legal documents when analysing water quality governance in China.

This section starts with the policy framework consisting of three Five-Year Plans (FYP) and their corresponding sub-plans. Then the discussion turns to the legal framework including concrete laws and regulations at national level. Implementation is briefly analysed in this chapter and further elaborated in Chapter 6.

3.1 Policy Framework of Water Quality Management in China

This section reviews the policy development of water quality management in China during the last decade. Key policies in three periods, the 11th FYP period (2006–2010), the 12th FYP period (2011–2015) and the 13th FYP period (2016–2020), are discussed. They set out the discourse for water development in China.

A. The period covered by the 11th Five-Year Plan

The 11th FYP is considered to be a milestone in the history of environmental protection in China. For the first time, it set binding goals for environmental management, i.e. applying total pollution load control to two major pollutants, sulphur dioxide (SO_2) (a main source of air pollution) and COD, respectively. Concrete reduction targets for those two pollutants are prescribed by its sub-plan, the 11th National Environment Protection Plan. The main approaches for achieving COD reduction include accelerating the construction of facilities for urban sewage treatment and recycling, and strengthening industrial wastewater treatment. According to the Plan, by the end of 2010, all cities must be equipped with sewage treatment facilities, and at least 70% of urban sewage should be treated. For controlling COD discharges from industry, the Plan focuses on the major industries, which account for 65% of China's total industrial COD wastewater discharges, but it also pays attention to the small, heavily polluting industries which have a low level of relevant equipment and poor environmental protection facilities ("The 11th Five-Year Plan for Economics and Social Development [2006–2010]", 2006).

Besides controlling COD, the other two tasks of the 11th FYP are addressing drinking water security and improving water pollution management in major river basins. In response to those two tasks, the former Ministry of Environmental Protection (MEP), jointly with other related ministries, published a National Plan of Urban Drinking Water Source Protection (2008–2020) and six prevention and control plans in major river basins (2006–2010) (Table 1). The former is the first national plan for urban drinking water source protection and applied to 4002 drinking water source areas nationwide. It identifies the main pollution sources for different levels of drinking water protection zones and provides corresponding management measures for pollution prevention and control. The overall goals and the indicators for evaluation are also prescribed by this national plan.

There are six pollution prevention and control plans which apply in six corresponding major river basins (Table 1). Each prescribes specific water quality targets and COD reduction targets. The COD targets in each plan are broken down and further assigned to provinces located within the basin in order to allocate clear responsibility (Sect. 4).

B. The period covered by the 12th Five-Year Plan

The 12th FYP went a step further towards environmental protection by adding two new binding pollution reduction targets for Ammonia Nitrogen and Nitrogen Oxides (a main source of air pollution), respectively ("The 12th Five-Year Plan for Economics and Social Development [2011–2015]", 2011). To achieve the reduction targets for COD and

Table 1 Policy framework of water quality management

FYP	*Document name*	*Water pollution-related targets or tasks*
11th FYP period	11th National Environment Protection Plan	• Achieving the reduction targets of COD and SO_2, respectively • Safeguarding the safety of surface drinking water sources • Enhancing water pollution prevention and control in major river basins
	National Plan for Urban Drinking Water Source Protection (2008–2020)	By 2020: • Fully controlling environmental pollution in urban drinking water source areas • Enhancing the capacity of emergency monitoring and water supply and • Comprehensively improving the water quality in urban drinking water source areas
	Water Pollution Prevention and Control Plan in Huai River, Hai River, Liao River, Chao Lake, Dianchi and Upper and Middle Reaches of the Yellow River (2006–2010)	• Setting water quality targets in corresponding river basins • Setting pollution control targets
12th FYP period	12th National Environment Protection Plan	• Achieving the reduction targets for COD, SO_2, Ammonia Nitrogen and Nitrogen Oxides, respectively • Strictly protecting drinking water sources • Enhancing water pollution prevention and control in major river basins • Paying attention to water pollution prevention and control in other river basins • Promoting groundwater pollution prevention and control • Enhancing rural environmental protection

(continued)

Table 1 (continued)

FYP	*Document name*	*Water pollution-related targets or tasks*
	National Plan for Groundwater Pollution Prevention and Control (2011–2020)	• Investigating groundwater pollution status • Safeguarding the safety of groundwater drinking water sources • Strictly controlling urban pollution which may contaminate groundwater quality • Strengthening pollution prevention control of major industries which may contaminate groundwater quality • Classifying and controlling agricultural diffuse pollution which may contaminate groundwater quality • Strengthening soil pollution prevention and control in order to protect groundwater quality • Carrying on groundwater restoration step by step • Developing a monitoring system for groundwater quality
	Water Pollution Prevention and Control in Major River Basins (2011–2015)	• Setting water quality targets in corresponding river basins • Setting total pollution load control targets for COD and Ammonia Nitrogen for industrial and domestic wastewater in corresponding river basins
	12th National Pollution Prevention and Control Plan for Livestock and Poultry Farming (2011–2015)	• Improving existing regulations, policies and standards of livestock and poultry farming • Promoting pollution abatement in livestock and poultry farming • Strengthening zoning management and pollution source control • Demonstrating and promoting techniques of water pollution prevention and control in livestock and poultry farming • Developing monitoring capacity in livestock and poultry farming
	No. 1 Document of 2011	Three Red Lines

(continued)

Table 1 (continued)

FYP	*Document name*	*Water pollution-related targets or tasks*
	Action Plan for Prevention and Control of Water Pollution	By 2020: • Periodically improving national water quality • Substantially reducing seriously polluted water bodies • Continuing to enhance the security level of drinking water • Strictly control over the exploitation of groundwater • preliminary containment of groundwater pollution and • Improving the water ecological environment in important regions By 2030: • Striving to improve the overall quality of national water quality • Restoring, on a preliminary basis, functioning system water ecology By 2050: • Comprehensively improving the national ecological environment
13th FYP period	The 13th Five-Year Eco-Environmental Protection Plan	By 2020: • More than 70% of surface water to be at Class III or above • Less than 5% of surface water to be at Class V • Water quality compliance rate of key water body function areas to reach 80% or above and • 15% of groundwater allowed to be in "very bad" category

Ammonia Nitrogen, besides continually improving urban sewage treatment facilities and strengthening water pollution management in major river basins, the main approaches in the 12th FYP included applying a system for controlling the total amount of pollutants discharged in the main polluted areas and polluting industries and intensive livestock and poultry farms.

The MEP published a comprehensive Pollution Prevention and Control Plan in Major River Basins (2011–2015) (Table 1), in which four additional basins (including major lakes and reservoirs) were added in comparison with the 11th FYP. The Plan divides the ten river basins into 36 Controlling Zones and 318 Controlling Units based on the natural characteristics of water catchment and the needs of administrative management. The reduction targets for COD and Ammonia Nitrogen are broken down into layers via the three-tier system of River Basin-Controlling Zone-Controlling Unit. The Controlling Units usually overlap with the lowest administrative level or are covered thereby. The Controlling Units are further divided into Priority Controlling Units and General Controlling Units; the water bodies in Priority Units are again divided into different types, namely, water bodies for preservation, pollution risk prevention or pollution control. Different measures are applied in different water bodies. The purpose of this system is to achieve meticulous management in river basins.

The discharge of COD and Ammonia Nitrogen contributed by livestock and poultry farms accounted for 45 and 25% of the total discharge amount, and 95 and 79% of the total agricultural discharge amount in 2010, respectively (Ministry of Environmental Protection & Ministry of Agricultrue, 2013). To cope with this, the MEP, jointly with the Ministry of Agriculture, published the 12th National Pollution Prevention and Control Plan for Livestock and Poultry Farms (2011–2015) (Table 1), in which the reduction targets for COD and Ammonia Nitrogen and the targets for waste treatment facilities in the livestock and poultry farms are explicitly formulated. Those targets are broken down and assigned to each province as well.

Groundwater gains more attention in the 12th FYP. This was stipulated under the category of water environment quality by the Plan, which has an equal status with the category of drinking water security, instead of ranking under it as in the 11th FYP. In response to this new situation, the MEP issued a National Plan of Groundwater Pollution Prevention and Control (2011–2020) (Table 1), which identifies the main challenges of groundwater pollution management and formulates both short-term and long-term targets, i.e. basically understanding the pollution situation and initiating restoration pilot schemes by 2015, and comprehensively monitoring typical pollution sources and enlarging the restoration scope by 2020. The main approaches include conducting investigations into pollution sources, strictly implementing environmental laws, and developing risk prevention mechanisms and so on.

A very influential policy in the 12th FYP period is the so-called "No. 1 Document of 2011", jointly issued by the Party's Central Committee and the State Council (The Central Committee of Communist Party & The State Council, 2011). It highlights the sustainable use of water resources over a ten-year period and sets the famous "Most Stringent Standards" which are usually called the "Three Red Lines", i.e. establishing a control limit for water resource exploitation and utilization, a control limit for water use efficiency, and a pollutant-carrying limit for water function zones. A year later the State Council published Guiding Opinions on the Implementation of the Most Stringent Water Management, which set up corresponding targets for the Three Red Lines. In terms of the pollutant-carrying limit, the discharged pollutants in water function zones must be controlled within the water pollutant-carrying capacity, and 95% of the water bodies must fulfil the required water quality standards by 2030. Targets are broken down and assigned to provinces (Sect. 4).

The most comprehensive plan, the Action Plan for Prevention and Control of Water Pollution, was issued by the State Council at the end of the 12th FYP period. The new plan, which includes a list of measures to tackle water pollution, forms part of China's efforts to combat all types of pollution. It follows a plan released in 2013 to tackle smog, which has gained initial success (Song, 2015). This water action plan aims to improve the quality of surface water and groundwater throughout the whole country by 2030. It is also called the "10-Point Water Plan" since it contains ten strategic measures as follows:

1. Taking overall control of pollutant discharges;
2. Promoting the transformation of the economic structure (to lower pollution intensity);
3. Focusing on the protection of water resources through saving water;
4. Strengthening support for science and technology;

5. Allowing market mechanisms to impact water and pollution levies;
6. Strengthening environmental law enforcement and supervision;
7. Strengthening the management of the overall hydrological cycle;
8. Ensuring the security of water for ecological and environmental purposes;
9. Confirming and implementing responsibilities for all parties involved; and
10. Strengthening public participation and social supervision (Han, Currell, & Cao, 2016).

Each measure prescribes corresponding approaches, indicators, scope and timeframes. For example, in order to achieve the purpose of taking overall control of pollutant discharges, small factories in sectors such as paper, insecticides and tanning which do not meet the national industrial policy must be shut down by the end of 2016; special pollution prevention and control programmes should be established for ten major industries including fertilizers and pesticides; industrial agglomeration areas should be equipped with centralized sewage treatment facilities and fitted with automatic online monitoring devices by the end of 2017; existing urban sewage treatment facilities should be altered according to local conditions and reach corresponding discharge standards or recycling requirements before the end of 2020, to name just a few.

This Action Plan is regarded as the blueprint for water quality management in the coming decade. It differs from previous approaches by stressing the philosophy of total water cycle health including groundwater, surface water and marine water, and acknowledging their interconnectedness. Furthermore, it aims to give rise to the development of an extensive environmental protection industry and an environmental protection services sector, in order to create a new area to boost sustainable economic growth (Han et al., 2016).

C. The period covered by the 13th Five-Year Plan

The environmental target of the 13th FYP (2016–2020) is to actualize an overall improvement of the country's environment by implementing the strictest possible environmental protection system and increasing the responsibility of polluters. According to the Plan, over 80% of the major rivers and lakes designated as water functional areas need to meet water quality standards, groundwater pollution surveys are required to be carried out and comprehensive prevention and treatment measures have to be in place. All industries are required to meet emissions standards, all heavily polluting enterprises located within urban districts will be either relocated, upgraded, or, in accordance with the law, shut down ("The 13th Five-Year Plan for Economics and Social Development [2016–2020]", 2016).

The 13th Five-Year Eco-Environmental Protection Plan, issued at the end of 2016, is the sub-Plan of the 13th FYP. The plan consists of three core missions: to raise the quality of the environment; to strengthen holistic management solutions; and to speed up the amendment of environmental issues. Water is again a key aspect of the plan. There are six general aims for improving the quality of China's water resources in the period of the Plan:

1. Establishing unit-based management of water resource quality;
2. Establishing holistic, basin-wide strategies to tackle pollution;
3. Priority protection of good quality water bodies;
4. Establishing holistic strategies to tackle groundwater pollution;
5. Strongly improving polluted urban water bodies; and
6. improving water quality of river mouth and nearshore areas (China Water Risk, 2016).

There are more defined targets related to water quality which are set out in the plan (Table 1). In general, this Plan is more stringent than the previous Plans.

3.2 *Legal Framework of Water Quality Management in China*

The legal framework of water quality management, under the umbrella of the Constitution and the Environmental Protection Law, consists of four categories which are divided according to their functions (Table 2). They are: natural resource laws; laws and regulations on pollution prevention and control; laws and regulations on environmental management; and normative documents such as environmental quality standards, emission standards and technical guidelines.

A. The Constitution and the Environmental Protection Law

The Chinese legal system is primarily based on the Civil Law model, and the primary source of law is statutes. The Constitution is at the top of legal system. It links together water pollution prevention and control as

Table 2 Legal framework of water quality management

Categories	*Laws, regulations, measures and standards*
Natural resource laws	Water Law, published in 1989, revised in 2002, amended in 2009, 2016 Water and Soil Conservation Law, published in 1991, revised in 2010 Fisheries Law, published in 1986, amended in 2000, 2004, 2009, 2013 Mineral Resources Law, published in 1986, amended in 1996, 2009 Grasslands Law, published in 1985, revised in 2002, amended in 2009, 2013 …
Laws and regulations on pollution prevention and control	Water Pollution Prevention and Control Law, published in 1984, amended in 1996, revised in 2008, 2016 Atmospheric Pollution Prevention and Control Law, published in 1987, amended in 1995, revised in 2000, 2015 Law on the Prevention and Control of Radioactive Pollution, published in 2003 Law on the Prevention and Control of Environmental Pollution Caused by Solid Wastes, published in 1995, revised in 2004, amended in 2013, 2015, 2016 Regulation on the Prevention and Control of Pollution from Large-scale Breeding of Livestock and Poultry, published in 2013 Regulation on Environmental Pollution from the Recycling of Ships, published in 1988, revised in 2016 …
Laws and regulations on environmental management	Law on an Environmental Impact Assessment, published in 2002, amended in 2016 Regulation on the Environmental Impact Assessment of Planning, published in 2009 Regulations on the Administration of Pesticides, published in 1997, revised in 2001 …

(continued)

Table 2 (continued)

Categories	*Laws, regulations, measures and standards*	
Normative documents	Water-related Environmental Quality Standards	Environmental Quality Standards for Surface Water GB 3838-2002
		Quality standards for groundwater GB/T14848-93
		Standards for Drinking Water Quality GB5749-2006
		Water Quality Standards for Fisheries GB 11607-89
		Standards for Irrigation Water Quality GB 5084-92
	Water-related Discharge and Emission Standards	Discharge Standards for Pollutants for Municipal Wastewater Treatment Plants GB 18918-2002
		Discharge Standards for Pollutants from Livestock and Poultry Breeding GB 18596-2001
		Emission Standards for Pollutants from the Petroleum Refining Industry GB 31570-2015
		Emission Standards for Pollutants from the Inorganic Chemical Industry GB 31573-2015
	Technical Guidelines	Technical Guidelines for an Environmental Impact Assessment General Provisions HJ 2.1-2011
		Technical Guidelines for the Environmental Impact Assessment of Surface Water Quality HJ/T 2.3-93
		Technical Guidelines for the Environmental Impact Assessment of Groundwater Quality HJ 610-2016

a responsibility for the state. "The state protects and improves the environment in which people live and the ecological environment. It prevents and controls pollution and other public hazards" (Article 26 of the Constitution 1982).

The Environmental Protection Law is the fundamental basic law in the field of environmental activities. The new Environmental Protection Law, which entered into force in 2015, sets out the principles for environmental protection, for example, the principle of giving priority to protection, focusing on prevention, comprehensive treatment, engaging the general public, and enforcing accountability for damage and so on. Specific to environmental pollution prevention and control, approaches prescribed in the Law include promoting clean production and the recycling of resources, implementing the "Three Synchronizations" (the facilities for pollution prevention and control must be conducted simultaneously with the main projects with regard to their design, construction and operation) (Dai, 2014), establishing a liability regime, establishing a pollution levy system, and implementing a total pollution load control system. This Law focuses on principles and the allocation of responsibility; more concrete regulations with regard to environmental protection are spread among the following different categories.

B. Natural resource laws

Natural resource laws are formulated for the purpose of the development and utilization of natural resources, although they also touch upon pollution prevention and control. For example, the purpose of the Water Law is laid down in Article 1, which states that "this law is formulated for the rational development, utilization, preservation, and protection of water, for the prevention and control of water disasters, and for the sustainable utilization of water resources in order to meet the needs of national economic and social development". Its scope of application lies "in the development, utilization, preservation, protection, and management of water resources and in the prevention and control of water disasters within the territory of China" (Article 2). Although it later deals with water pollution prevention, for example in Article 34 which states that the "construction of any outlet for sewage discharge in the protection zones of drinking water sources is prohibited", it explicitly states that "water pollution prevention and control shall be implemented in accordance with Water Pollution Prevention and Control Law" (Article 81).

Other laws in this category share the same character in that they mainly focus on the development and utilization of corresponding natural resources. For example, the Water and Soil Conservation Law is intended "to prevent and control water and soil loss, protect and reasonably utilize water and soil resources, reduce flood disasters, drought and sandstorms", while the Fisheries Law has been formulated "for the purpose of enhancing the protection, increase, development and reasonable utilization of fishery resources" and the Grasslands Law is intended "to protect, develop and make rational use of grasslands".

Although conservation and sustainable utilization could indirectly contribute to the improvement of environmental quality, this is not really the focus of this category of laws.

C. Laws and regulations on pollution prevention and control

The main purpose or responsibility of this category is to prevent and control environmental pollution. The most relevant law for water quality management is the Water Pollution Prevention and Control Law, which has been formulated for the purpose of "preventing and controlling water pollution, protecting and improving the environment, safeguarding drinking water safety…". It applies to "rivers, lakes, canals, irrigation channels, reservoirs and other surface waters and groundwater". The law regulates important aspects of water quality management, including the policymaking bodies for water quality standards and discharge standards, the supervision bodies for water pollution prevention and control activities, and both general and specific measures for preventing and controlling industrial pollution, urban domestic pollution, agricultural diffuse pollution, pollution from ships and the pollution of drinking water sources and other special water areas such as scenic areas and important fishery water bodies.

The Water Pollution Prevention and Control Law was newly revised in 2017. In comparison with the previous version, 11 new articles have been added. It enhances the supervision of trans-provincial river basin pollution, increasing penalties for water pollution which exceeds emission standards, extending the scope of the application of pollutant discharge permits, and strengthening pollution source control and so on, which are all existing barriers to effectively managing water quality in practice.

Other laws and regulations in this category, although they do not directly manage water pollution, nevertheless contribute to water quality management since water is closely interconnected with many other environmental activities. Combating one type of pollution often produces gains as regards another form of pollution. For example, the implementation of the Law on the Prevention and Control of Environmental Pollution Caused by Solid Wastes could directly contribute to the prevention of water pollution in related water bodies since, according to the law, "no entity or individual may dump solid wastes into or pile them up in rivers, lakes, ditches, reservoirs, bottomlands, banks or slopes under the highest waterline …". Similarly, the implementation of the Regulation on the Prevention and Control of Pollution from Large-scale Breeding of Livestock and Poultry could also prevent water pollution, as the Regulation explicitly requires that "livestock husbandry development plans shall take the environmental carrying capacity into consideration" and "setting up livestock and poultry farms and breeding areas in protection zones of drinking water sources is forbidden".

D. Laws and regulations concerning environmental management

The third category consists of laws and regulations that relate to environmental management. The Law on an Environmental Impact Assessment (EIA) is one of the most important laws in this regard. It was formulated to prevent the unfavourable impacts of programmes and construction projects upon the environment after they are carried out. Together with its technical guidelines for an environmental impact assessment in surface water and groundwater (the fourth category), the Law provides for principles, working procedures, working methods and requirements for a water quality assessment of possible impacts on surface water and groundwater caused by construction projects.

Another notable source of regulation in this category are the Regulations on the Administration of Pesticides. China is the largest consumer of pesticides although it has only 7% of the world's arable land. The annual usage is 2.5 times higher than the global average. The misuse of pesticides, as well as fertilizers, has caused serious water pollution (Dai, 2014). The Regulations on the Administration of Pesticides and their Implementation Measures regulate the supervision and administration of the production, marketing and application of pesticides. To a large extent they standardize the use of pesticides at the national level, and if they are effectively implemented they will be of assistance in reducing diffuse water pollution to a certain extent. The overuse of fertilizers is another trigger for water pollution and the formulation of a regulation for fertilizer usage has been proposed for many years; however, such a regulation has not yet been given its final shape.

E. Normative documents on water quality management

This category contains a number of normative documents and three sub-categories are included therein. They are the water-related environmental quality standards, the water-related emission standards and related technical guidelines (Table 1). These normative standards and guidelines provide a scientific basis for environment-related decision-making such as the formulation of environmental plans and discharge reduction targets.

The Water-related Environmental Quality Standards stipulate indicators and indexes for water quality in different water bodies. The Environmental Quality Standards for Surface Water (GB 3838-2002) define five classes for water bodies according to their use. Some 109 parameters, including 24 basic parameters, 5 supplementary parameters, and 80 specific parameters are prescribed by the Standards. According to these Standards, those basic parameters are implemented for rivers, lakes, canals, channels, and reservoirs, while supplementary and specific indicators are applied to drinking water resources, and specific indicators could be determined and selected by local governments according to local environmental characteristics and protection requirements (Su et al., 2017). These parameters provide clear references for establishing water targets and water quality monitoring in daily life.

The water-related emission standards define quantitative limits for wastewater discharges from specific sources over specific timeframes. A total of 46 national discharge and emission standards have been published for a wide range of industries since the 12th FYP, ranging from chemicals to coal to pharmaceuticals and to sugar. These standards play a very important role in water quality monitoring and assessment. Together with the above Environmental Quality Standards, they work as yardsticks for judging whether the existing regulations have been adhered to. Those who violate these standards must bear corresponding legal responsibility.

The technical guidelines provide a standard code for implementing the related regulations. For example, the Technical Guidelines for the Environmental Impact Assessment of Surface Water Quality

(HJ/T 2.3–93) provide for terminology, standard symbols, working procedures, and evaluation methods for the environmental impact assessment in different surface water bodies. They standardize operational norms in this regard and have practical significance for water environment management.

4 Implementation

The central government has striven to improve the country's water quality during the past decade, and some positive results have been achieved. By the end of the 11th FYP period, according to the 12th FYP, the reduction targets for SO_2 and COD had been exceeded. The treatment rate of urban sewage had increased to 72% in 2011, as compared with 52% in 2005. The 12th FYP targets for the reduction of SO_2, COD, Ammonia Nitrogen and Nitrogen Oxides had also been exceeded. In 2015 the average concentration of COD in surface water had fallen by 32% when compared with 2010, which has directly contributed to the improvement of water quality. The quality of surface water in water bodies evaluated nationwide in 2015 classified as better than Class III was 14.6% higher than in 2010, and the inferior Class V was 6.8% lower (The State Council, 2016).

However, the decline in pollutants without binding reduction targets, like TP, has been at a much lower rate in comparison with the above four pollutants with binding reduction targets (The State Council, 2016). Whether or not the targets are binding plays a key role in the success or failure to meet the necessary reductions. In fact, around 80% of the COD reduction target had not been accomplished during the 10th FYP period as it was only a recommended target at that time. The binding targets are implemented by the Government Objective Responsibility System, which is a core operating tool in environmental management in China (see Chapter 6). The central government establishes binding targets for key areas of development priority at the national level. Through the Government Objective Responsibility System, the overall established targets are assigned to provincial governments, which then reassign their quotas to lower levels (municipal and county levels). From there the quotas are further allocated to local major polluting enterprises or to wastewater treatment plants and so on. Party leaders and government officials are required to sign responsibility contracts with the upper level and their performance will be assessed by the latter. Officials who fail to meet the binding targets will be held accountable (Chapter 6).

5 Conclusion

China has been witnessing continuous high economic growth during the last decade. The rapid economic growth and urbanization are ushering in a new phase of continuous pressure on the environment. Traditional pollution management schemes are not able to meet these challenges. In order to adapt to the new situation, many efforts have been made by the central government. A number of new instruments have been developed to tackle different water pollutants and pollution sources. However, there are still some concerns with respect to the implementation of those instruments.

For example, there is too much focus on binding targets for particular pollutants and this leads to the prevention and control of other important pollutants being neglected. This "one falls, another rises" effect could ultimately result in a failure to improve water quality in general. Furthermore, the improvement in water quality during the last decade has actually largely relied on, besides the implementation of binding reduction targets, engineering measures such as improving the facilities in wastewater treatment plants or livestock and poultry farms. Soft measures such as cooperation and coordination between different sectors have not been extensively developed, which is still one of the main obstacles in general water management in practice. Implementation strongly relies on governmental administration; the judicial system has lagged behind to a large extent.

The Action Plan for the Prevention and Control of Water Pollution and the 13th Five-Year Eco-Environmental Protection Plan offer a turning point with respect to water pollution management. However, they are more like blueprints at the current stage; more concrete implementation measures are expected to be developed in the coming years.

References

China Water Risk. (2016). *China's 13th Five-Year Plan for Ecological & Environmental Protection (2016–2020)*. Retrieved from http://chinawater-risk.org/notices/chinas-13th-five-year-plan-2016-2020/.

Dai, L. (2014). Something Old, Something New, Something Borrowed and Something Blue—Tackling Diffuse Water Pollution from Agriculture in China: Drawing Inspiration from the European Union. *Utrecht Law Review, 10*(2), 136–154.

Dai, L. (2015). Regulating Water Pollution in China and the European Union with a Focus on Agricultural Pollution. *Journal of Water Law, 24*(3/4), 150–156.

Hajer, M. (1993). *Discourse Coalitions and the Institutionalisation of Practice: The Case of Acid Rain in Great Britain* (pp. 43–67). Durham, NC and London: Duke University Press.

Han, D., Currell, M. J., & Cao, G. (2016). Deep Challenges for China's War on Water Pollution. *Environmental Pollution, 218*, 1222–1233.

Jeffares, S., Held, G., & Atkinson, R. (2010). Theories of Discourse and Narrative: What Do They Mean for Governance and Policy? In *Sustainability in European Environmental Policy* (pp. 137–152). Abingdon: Routledge.

Ministry of Environmental Protection. (2017). *2016 China Environmental Situation Bulletin*. Beijing. Retrieved from http://www.mep.gov.cn/gkml/hbb/qt/201706/W020170605812243090317.pdf.

Ministry of Environmental Protection & Ministry of Agriculture. (2013). *The 12th National Pollution Prevention and Control Plan for Livestock and Poultry Farms (2011–2015)*. Beijing.

Mundial, B. (2007). *Cost of Pollution in China: Economic Estimates of Physical Damages*. Banco Mundial.

Song, M. (2015). *China Headlines: Water Pollution Plan to Bring Long-Term Gain*. Retrieved from http://www.xinhuanet.com/english/2015-04/17/c_134160737.htm.

Su, J., Ji, D., Lin, M., Chen, Y., Sun, Y., Huo, S., … Xi, B. (2017). Developing Surface Water Quality Standards in China. *Resources, Conservation and Recycling, 117*, 294–303.

The 11th Five-Year Plan for Economics and Social Development (2006–2010), (2006).

The 12th Five-Year Plan for Economics and Social Development (2011–2015), (2011).

The 13th Five-Year Plan for Economics and Social Development (2016–2020), (2016).

The Central Committee of Communist Party & The State Council. (2011). *Decision on Promoting Water Conservancy Reform and Development*.

The State Council. (2016). *Report on the Environmental Situation and the Implementation of Environmental Targets in 2015*. Retrieved from http://www.npc.gov.cn/wxzl/gongbao/2016-07/11/content_1994454.htm.

van Rijswick, H. F., Edelenbos, J., Hellegers, P., Kok, M., & Kuks, S. (2014). Ten Building Blocks for Sustainable Water Governance: An Integrated Method to Assess the Governance of Water. *Water International, 39*(5). https://doi.org/10.1080/02508060.2014.951828.

Xie, J. (2008). *Addressing China's Water Scarcity: A Synthesis of Recommendations for Selected Water Resource Management Issues*. Washington, DC: The World Bank.

CHAPTER 3

Public Participation in Water Governance in China

Abstract Water governance can be considered complex processes in which various actors with different interests discuss, deliberate and negotiate problem analysis and solution-finding to water issues. For a country without a mature representative democracy like China, the development of public participation mechanisms is of great importance. This chapter focuses on public participation. It discusses whether China's current regulations have offered the public opportunities to participate. It not only looks exclusively at public participation in China's water law, but also examines public participation obligations in environmental law which has a bearing on the governance of China's freshwater resources.

Keywords China · Water governance · Public participation · Public-interest litigation

1 Introduction

Water governance can be considered as complex processes in which various actors, with different and often opposing values, viewpoints and interests, discuss, deliberate and negotiate problem analysis and solution-finding to water (van Rijswick, Edelenbos, Hellegers, Kok, & Kuks, 2014). This multi-actor complexity means taking into account different perceptions with respect to the problems, interests and positions of the stakeholders, which can be addressed by engaging stakeholders

L. Dai, *Politics and Governance in Water Pollution Prevention in China*, Politics and Development of Contemporary China,
https://doi.org/10.1007/978-3-030-02874-9_3

across the water chain, related sectors and levels of government in an iterative dialogue process (Akhmouch & Clavreul, 2016). For a country like China without a mature representative democracy, the development of public participation mechanisms is of great importance.

Public participation has been described as "a process by which the views and concerns of the public are identified and incorporated into decision making by public institutions" (Meganck & Saunier, 2012). According to Smith (1983), public participation encompasses a range of procedures and methods designed to consult, involve and inform local communities and citizens (i.e. the "public") (Akhmouch & Clavreul, 2016). There are in fact many definitions of "public participation" (IAP2; Ramlogan, 2010; Rowe & Frewer, 2004). They all depart from the same idea, namely that institutions should allow people with an interest in a decision-making process to be involved in some way in that process.

In the water sector, the "public" essentially encompasses civil society and customers (Akhmouch & Clavreul, 2016). Another term which needs to be distinguished from public participation is "stakeholder engagement", which goes beyond civil society and end users, and reaches out to other groups of actors within and outside the water sector in activities related to planning, decision making, implementation, and monitoring and evaluation. Stakeholder engagement encompasses different levels of governments, private sectors, regulators, service providers, donor agencies, investors, civil society in its different forms (e.g. citizens, non-governmental organizations, users' movements, etc.) and other relevant constituencies (Akhmouch & Clavreul, 2016). Thus, public participation can be seen as a practice of stakeholder engagement.

This chapter only focuses on public participation. It does not describe or study the governance systems per se but focuses on whether China's current regulations have offered the public opportunities to participate. Furthermore, this chapter does not look exclusively at public participation in China's water law; it also examines public participation obligations in environmental law which have a bearing on the governance of China's freshwater resources.

2 Public Participation in China's Water-Related Regulatory Framework

Before turning to the analysis, it is perhaps helpful to say a few words about the methodology and the sources used in this section. China's law is rather complex. At the top, there is the Constitution of 1982, recently amended in 2018. This Constitution says little to nothing about public participation except that "All power in the People's Republic of China belongs to the people" (Article 2 of the Constitution 1982). Then there are the basic laws, promulgated by the National People's Congress (NPC); administrative regulations, promulgated by the State Council; local regulations promulgated by the local People's Congress; and administrative rules including local and departmental rules.

The Five-Year Plans (FYP), although not legal documents, do set binding targets and, as stated in the previous chapter, they offer a glimpse of the priorities to which the central and local governments will adhere in the next five years. They are analysed first because they set out the general strategy, which is then translated by the legislature into concrete laws and regulations. When it comes to sectoral regulations, the former Ministry of Environmental Protection (MEP, which has been merged with the Ministry of Ecological Environment in 2018) is the focal point for water pollution management. Therefore, the ministerial regulations made by the MEP are also analysed.

Based on the above explanation, this section first examines the three most recent FYPs, then the basic laws, national administrative regulations, and ministerial regulations.

2.1 Public Participation in China's Five-Year Plans

The 11th FYP (2006–2010) stipulated that the public should "be encouraged to supervise the environmental protection activities and participate". The Plan has been considered to be a turning point in China's environmental protection policies, since it is no longer just economic in focus but also gives attention to environmental protection, for example, by setting up binding environmental targets (Chapter 1). Those binding targets are usually considered as "hard goals", which are formulated in clear terms and with specific deadlines, and the results are usually detailed and precise. In other words, attaining these goals can be considered an obligation of result. The targets, which are not binding in the Plans, are usually taken as "soft goals", i.e. they are regarded as "recommendations" (Dai, 2015).

The above-mentioned principle of public participation in the 11th FYP is considered to be a "soft goal", or "recommendation", since the consequence of a failure to attain the goal of public participation, as determined by the Plan, is not specific, and neither is there an accompanying deadline. The Plan calls for "improving the scientific and democratic decision-making mechanism, and the collective decision-making on major issues, expert advice, public notices and hearings, and decision-making accountability system [...] protecting the citizens' right of access to information, participation, expression and supervision on governmental work" ("The 11th Five-Year Plan for Economics and Social Development (2006–2010)", 2006).

The 12th FYP (2011–2015) strengthens environmental protection by adding more binding targets (see Chapter 1 above) but, regarding public participation, the statement is similar to the 11th FYP. The relevant departments are urged to "listen to public opinion, through public hearings, when formulating environmental standards" ("The 12th Five-Year Plan for Economics and Social Development (2011–2015)", 2011).

The 13th FYP (2016–2020), in the section on "reform of fundamental environmental governance systems", states that the government will "strengthen the public disclosure of environmental information, improve the channels for public participation in pollution control efforts, and improve the environmental public-interest litigation system". In another section, that on "public participation", it is stated that the government will "in accordance with the law, safeguard residents' rights to be informed, to participate, to make decisions, and to conduct oversight, and improve institutional channels for public participation in governance. For major decisions related to key public interests, [the government] will broadly solicit comments and suggestions from the public through residents' meetings, discussions and deliberations, and democratic hearings, and strengthen public scrutiny and assessment by improving democratic evaluations and the transparency of village and community affairs" ("The 13th Five-Year Plan for Economics and Social Development (2016–2020)", 2016).

Within the period of the three FYPs, the State Council issued many supporting policies, the most noteworthy being the Action Plan for Prevention and Control of Water Pollution. It is the toughest and most comprehensive plan to date and has been discussed in Chapter 1. It was issued by the State Council at the end of the 12th FYP period. One of the highlights of the Plan is its emphasis on public participation. It is the first time that the government has enlisted ordinary citizens to help fight its "war on pollution" (Blanchard & Stanway, 2014). In the section on "enhancing public participation and social monitoring", the Plan sets out a number of measures to strengthen public participation. For example, it requires the governments at all levels to make environmental information public; provide training and consultation to the public and to social organizations on laws and regulations on water pollution prevention and control; and to strengthen publicity and education by including knowledge of water resources, water environment protection and the water regime in the national education system. It requires governments to establish a sound reporting system and give full backing to the "12369" environmental protection reporting hotline and network platform, and to solve environmental complaints reported by the public within a deadline.

The Plan pledges to the public that the "black and odorous water" (heavily polluted water) in urban areas will be controlled within 10% of the waterbodies by 2020. In response to this, the former MEP opened a public account on We Chat, an extremely popular Chinese app similar to Twitter, where members of the public can submit pictures and descriptions of any waterbody which they believe should be designated as black and odorous water. The former MEP guarantees that each report will receive an official response in seven working days or less. This has proved valuable for water quality monitoring. To date, there have been 2100 waterbodies reported by members of the public which have been identified as "black and odorous water", of which 1120 have been treated (Ministry of Housing and Urban-Rural Development & Ministry of Ecology and Environment, 2018). The platform also publishes weekly Reports, which take stock of the updated status of the "black and odorous water", as well as the formulation of clean-up plans and progress towards completion for the public.

Another landmark move was the Fourth Plenary Session of the 18th Central Committee of the Communist Party of China (CCCPC) in 2014, when the CCCPC's Decision on Major Issues Concerning Comprehensively Pushing Forward the Rule of Law was published. The Decision is instrumental in developing a transparent government, safeguarding democratic rights and the formation of civil society, and constructing and implementing systems for public participation and information disclosure (Fang, Wei, & Logan, 2017). Specifically, it promises to widen channels for the public to participate in legislative work in an orderly manner through discussion, hearing, assessment and publicizing draft laws.

2.2 *Public Participation in the Basic Laws*

The basic laws are issued by the (Standing Committee of the) National People's Congress in China. They have the highest legal standing in the country, second only to the Constitution. In 2014, the Environmental

Protection Law was revised for the first time since its enactment in 1989, the original 47 articles have been expanded to 70 articles.

From 2011 to 2014, the Standing Committee of the National People's Congress went through reviews of draft amendments to the old Environmental Protection Law before adopting the amendment in April 2014. The revision process witnessed an unprecedented move from the National People's Congress with a two-stage disclosure and comments process. The draft bill was disclosed for public comments, deliberated by the lawmakers and then disclosed again for a second round of public comments (Zhu & Wu, 2017). This move marks a significant change in public participation in China's environmental lawmaking processes.

The new revised law identifies public participation as one of its basic principles. It added a whole chapter to prescribe "information disclosure and public participation", albeit with a focus on information disclosure. According to the new law, governmental agencies that are responsible for or involved in environmental protection must publicize the information and data on environmental quality, management and supervision, etc. Evaluations of the environmental performances of both local environmental agencies (and their responsible officials, see Chapter 6) and of industries are required to be published as well.

The new law entailed a significant step towards public participation by prescribing rules on public interest litigation. Social organizations which satisfied two criteria could file lawsuits. Those criteria are that the organization should: (1) be registered with a government civil affairs department at or above the level of a city with districts; and (2) be engaged specifically in public service activities in environmental protection for five consecutive years without any record of violation of laws. The law also states that public interest litigants may not file suits to seek economic benefits. Profound changes have occurred in the landscape of litigation on environmental protection since the law was enacted in 2015. Dozens of public interest litigation suits were filed in the first month of 2015 (Zhang & Mayer, 2017). According to a press release from the Environment and Resources Division of the Supreme People's Court in March 2017, Chinese courts accepted 112 public interest litigation suits filed by non-governmental organizations (NGOs) in 2015 and 2016—a sharp increase from around eight cases per year before 2015 in the provinces and cities with a trial programme (Zhang & Mayer, 2017). One-quarter of the litigation suits were related to water pollution.

Under the umbrella of the Environmental Protection Law, there are environment-related basic laws, which could be further divided into three categories based on their functions, as discussed in Chapter 2: pollution prevention and control laws; resource conservation and utilization laws; and environmental management laws. Clearly, all three categories have a link with the governance of freshwater resources.

Many of the laws in the first group—pollution prevention and control—touch upon public participation. For example, the Air Pollution Prevention and Control Law requires that, "while making or revising the air environmental quality standards and the air pollutants emission standards, the opinions of [...] the general public shall be solicited" (Article 10). Article 32 of the Water Pollution Prevention and Control Law prescribes that "any enterprise or public institution or any other producer or dealer that discharges poisonous and hazardous water pollutants ... shall release the information on poisonous and hazardous pollutants..." (though this is only about information disclosure). The Law on the Prevention and Control of Environmental Pollution by Solid Waste and the Law on Prevention and Control of Radioactive Pollution also proclaim the same right.

The laws in the second category—resource conservation and utilization laws—have in common with the laws of the first group that the principle of public participation is only touched upon in a general way. For example, the Water and Soil Conservation Law requires that "while making water and soil conservation plans, the opinions of experts and the general public shall be solicited" (Article 13). Similarly, the Renewable Energy Law states that "when developing a plan for the development and utilization of renewable energy resources, [...] the department responsible for developing the plan shall solicit comments from relevant entities, experts and the general public" (Article 9). Other laws in this group, including the Water Law, the Agricultural Law, the Forest Law, the Mineral Resource Law and the Marine Environmental Protection Law do not refer to the principle at all.

The last category of basic laws contains those laws that relate to environmental management. The most relevant law in this category is the Environment Impact Assessment Law (EIA). The EIA Law makes public participation a required component of the process of environmental assessment. It requires that "the State encourages relevant entities, experts and the general public to participate in the appraisal of the environmental impacts in appropriate ways" (Article 5). The EIA Law requires an environmental impact assessment only for Governmental Plans and Construction Projects. Governmental Plans can relate to land use and regional, watersheds and sea areas development, as well as plans for industry, agriculture, energy, transportation, urban construction, tourism, and natural resource development. For Governmental Plans, the public participation rule is as follows:

> In the case of a programme may cause an unfavourable environmental impact or directly involve the environmental interests of the general public, the body that carries out the special programmes shall, prior to submitting the draft of the programmes for examination and approval, seek the opinions of the relevant entities, experts and the general public [...] by holding demonstration meetings or hearings or by any other means, except [when] it is provided by the state that a programme shall be kept confidential. (Article 11)

In the case of Construction Projects, the exact obligation depends on the "seriousness" of the potential environmental impact which the project is anticipated to have. When a Construction Project has a "significant" potential environmental impact, the developers must prepare a full-fledged environmental impact assessment report. When a project is anticipated to have a "light" potential environmental impact, the developers are only required to fill in an environmental impact report form. And for projects with a "very small" environmental impact, developers need only submit an environmental impact registration form (Article 16). Only the projects with a "significant" potential environmental impact demand public participation. In such cases, the construction entity "shall, before submitting the construction project for examination and approval, seek the opinions of relevant entities, experts and the general public by holding demonstration meetings, hearings or by any other means" (Article 21). Therefore, although public participation is compulsory in the EIA Law, the scope of application is very narrow. It is important to note that the measures to implement the principle of public participation, as prescribed by the Environmental Protection Law, are regulated by the former MEP in the form of ministerial regulations and will be discussed in detail below.

2.3 Public Participation in the National Administrative Regulations

The national administrative regulations are issued by the State Council. There are two different categories of regulations, i.e. the general administrative regulations and the trial regulations. Those related to public participation are analysed below.

The first category, the general administrative regulations, is aimed at managing governmental action. The most relevant regulation in this category is the general Regulation on Disclosure of Government Information enacted in 2007. It is established, inter alia, "for the purpose of safeguarding the legal access to government information by citizens, legal persons and other organizations [and] improving the transparency of government work" (Article 1). It is widely considered as a milestone in the country's information policy history and praised as a "sunshine programme" (Zhang, Mol, He, & Lu, 2010). The MEP was the first ministry to operationalize these general regulations in Measures for the Disclosure of Environmental Information (for Trial Implementation). The Regulation on Disclosure of Government Information provides supporting methods to implement the principle of public participation by regulating the scope of disclosure of governmental information, as well as the methods, the procedure, the monitoring, and the safeguarding methods. In terms of scope, two types of information disclosure were prescribed by the regulation, i.e. the voluntary information disclosure by the government and the information disclosure requested by public application.

The second category of administrative regulations is the category of administrative regulations that are tested as such and, when they function well, will be promoted to basic laws. They actually share the same characteristics with the basic laws in that they only make general provisions on the principle of public participation. For example, the Regulation on Nature Reserves states that "all units and individuals shall have [...] the right to inform against or lodge complaints with the units or individuals who damage the nature reserves" (Article 7).

2.4 *Public Participation in the Ministerial Regulations*

The Ministerial Regulations are issued by ministerial-level departments for matters within their area of responsibility. This article focuses on the regulations made by the MEP, since environmental protection and water pollution prevention in China is the responsibility of that ministry. The MEP must develop national environmental protection policies and plans, thereby taking the lead in environmental water governance.

Pursuant to the basic laws of the EIA Law enacted in 2003, the MEP issued the Interim Measures for Hearing the Administrative Licence in Respect of Environmental Protection (2004) to regulate the administrative licence in the field of environmental protection. These Interim Measures prescribe the conditions for public hearings. They stipulate that "the administrative department that organizes a [public] hearing shall abide by the principles of openness, fairness, impartiality and convenience-for-people, fully hear the opinions from citizens, legal persons and other organizations, and guarantee their respective rights to state their respective opinions, to cross-examine, and to defend themselves" (Article 4). The scope of the hearing includes the implementation of an administrative licence in respect of environmental protection; those construction projects that may have a serious influence upon the environmental quality of residents' living conditions; and any special planning which may have a negative influence upon the environment and directly affect the environmental interests of the public. The Interim Measures stipulate the rights and duties of the host and audience of the hearing, the procedure, and the safeguarding measures, as well as the standard forms to be used at the hearings.

Another regulation, the Interim Measures for Public Participation on Environmental Impact Assessment, prescribes a number of similar principles which applied in the case of the administrative licence as stated above. Again, it is stipulated that public participation shall be "open, fair, extensive and convenient-for-public" (Article 4). These Interim Measures are exclusively applied in relation to environmental impact assessments. It is stipulated that, "after the information disclosure, the public shall be solicited via surveys, consultations, seminars, debates, or hearings" (Article 12). The MEP also issued Technical Guidelines for Environmental Impact Assessment.

The Measures for the Disclosure of Environmental Information (for Trial Implementation), enacted by the MEP in 2008, are designed to implement the general Regulation on Disclosure of Government Information in the field of environmental protection. The trial Regulation requires not only environmental authorities but also industries to disclose environmental information. It groups the environmental information into four major categories: environmental laws and regulations; environmental quality; environmental management and supervision; and environmental accidents and emergency responses (Zhang et al., 2010). It requires the establishment of monitoring, evaluation and supervision systems. Those environmental authorities violating the rules of information disclosure will be held responsible and accountable by their supervisors; and industries violating the rules will face penalties.

To establish and improve the system of supervisory monitoring and information disclosure, the MEP issued two Measures in 2013, i.e. (1) the Measures on Supervisory Monitoring and Information Disclosure of Pollution Sources from Key State-Monitored Enterprises (for Trial Implementation) and (2) the Measures on Self-Monitoring and Information Disclosure of Pollution Sources from Key State-Monitored Enterprises (for Trial Implementation). Both took effect on 1 January 2014. These measures require, for the first time, that key state-monitored enterprises release real-time data on bi-hourly wastewater emissions, which must be publicly disclosed on online platforms of local, provincial-level environmental protection authorities (Fang et al., 2017). Although the implementation of these measures has, to a large extent, improved the public's access to environmental information from enterprises across multiple, heavily polluting industries, the effects are less than satisfactory in practice since there is a lack of an independent third party to verify the environmental information from the enterprises, and incentive mechanisms for rewards and punishment to support the implementation are lacking in the Measures as well. Therefore, in many cases, enterprises were willing only to disclose information advantageous to their business (Fang et al., 2017).

In 2015, the MEP enacted Measures for Public Participation in Environmental Protection, for the purpose of "safeguarding the legal rights of citizens, legal persons, and other organizations to obtain environmental information, supervise and participate in the environmental protection" (Article 1). The Measures define the subject matter and scope of public participation, which includes input from citizens, legal entities and other organizations. Specifically, the Measures call for public participation in the process of formulation of environmental policies and regulations, implementation of administrative licencing and imposition of administrative penalties, supervision of illegal acts, and implementation of publicity and education. The Measures indicate that "the environmental protection departments could solicit the public opinions via public reviews, questionnaires, workshops, expert meetings, or hearings; the public, legal persons and other organizations could provide comments and suggestions via telephones, letters, faxes, or internet" (Article 4). The Measures also, in very general terms, describe the procedures for soliciting the public input for the MEP and other environmental protection departments.

3 Conclusion

This chapter examines China's public policies and legal norms at the national level to explore whether those policies and norms offer the public opportunities to participate in water affairs. It finds that the central government has expanded public participation in all processes, for example, by promoting it in national strategic plans and strengthening it in newly revised laws. A skeleton structure for public participation has been basically built at the national level, though still with deficiencies. Obstacles to the effectiveness of environmental public participation still remain in practice. This is because, historically, China's governance is subject to a lack of public participation, "a wealth of laws with shallow roots", as one scholar once described it (Wu, 2008). A suggestion would be to apply the existing norms more rigorously by closer supervision and, parallel to this, also strengthening general legal awareness and education, which are fundamental for the implementation of public participation.

References

Akhmouch, A., & Clavreul, D. (2016). Stakeholder Engagement for Inclusive Water Governance: "Practicing What We Preach" with the OECD Water Governance Initiative. *Water, 8*(5), 204.

Blanchard, B., & Stanway, D. (2014). *China to 'Declare War' On Pollution, Premier Says.* Retrieved from https://www.reuters.com/article/us-china-parliament-pollution/china-to-declare-war-on-pollution-premier-says-idUSBREA2405W20140305.

Dai, L. (2015). A New Perspective On Water Governance in China: Captain of the River. *Water International, 40*(1), 87–99. https://doi.org/10.1080/02508060.2014.986702.

Fang, K., Wei, Q., & Logan, K. K. (2017). Protecting the Public's Environmental Right-to-Know: Developments and Challenges in China's Legislative System for EEID, 2007–2015. *Journal of Environmental Law, 29*(2), 285–315.

IAP2. Advancing the Practice of Public Participation. *IAP2 International Association for Public Participation.* Retrieved from http://www.iap2.org/search/all.asp?bst=core+values. Accessed November 11, 2018.

Meganck, R. A., & Saunier, R. E. (2012). *Dictionary and Introduction to Global Environmental Governance.* London: Routledge.

Ministry of Housing and Urban-Rural Development & Ministry of Ecology and Environment. (2018). *Information Disclosure of Urban Black and Odorous Water.* Retrieved from http://www.hcstzz.com/.

Ramlogan, R. (2010). *Sustainable Development: Towards a Judicial Interpretation* (Vol. 9). Leiden: Martinus Nijhoff Publishers.
Rowe, G., & Frewer, L. J. (2004). Evaluating Public-Participation Exercises: A Research Agenda. *Science, Technology and Human Values, 29*(4), 512–556.
Smith, D. H. (1983). Synanthrometrics: On Progress in the Development of a General Theory of Voluntary Action and Citizen Participation. *International Perspectives on Voluntary Action Research*, 80–94.
The 11th Five-Year Plan for Economics and Social Development (2006–2010), (2006).
The 12th Five-Year Plan for Economics and Social Development (2011–2015), (2011).
The 13th Five-Year Plan for Economics and Social Development (2016–2020), (2016).
van Rijswick, H. F., Edelenbos, J., Hellegers, P., Kok, M., & Kuks, S. (2014). Ten Building Blocks for Sustainable Water Governance: An Integrated Method to Assess the Governance of Water. *Water International, 39*(5). https://doi.org/10.1080/02508060.2014.951828.
Wu, J. (2008). Public Participation in the Enforcement of China's Anti-pollution Laws. *Law Environment and Development Journal, 4* 35–48.
Zhang, L., Mol, A. P., He, G., & Lu, Y. (2010). An Implementation Assessment of China's Environmental Information Disclosure Decree. *Journal of Environmental Sciences, 22*(10), 1649–1656.
Zhang, Q., & Mayer, B. (2017). *Public Interest Environmental Litigation Under China's Environmental Protection Law, 1*(2), 202–228.
Zhu, X., & Wu, K. (2017). Public Participation in China's Environmental Lawmaking: In Pursuit of Better Environmental Democracy. *Journal of Environmental Law, 29*(3), 389–416.

CHAPTER 4

China's Path to a Green Water Law

Abstract China's economic performance during the reform period is sometimes described as a miracle. However, the rapid economic growth has exacted an enormous environmental cost. The Chinese government has realized that a new, more sustainable development pathway is needed, which associates with a more sustainable development. This chapter looks at this recent shift which towards a "greener" Chinese water law. It analyzes in what way a sustainable development friendly approach to the utilization and management of water resources is reflected in China's domestic water law. Two themes, sustainable development and ecosystems approach, as well as how they are resurfaced on China's domestic water law are discussed in this chapter.

Keywords China · Sustainable development · Ecosystem approach · Water law

1 Introduction

China's economic performance during the reform period (since 1978 to now) is sometimes described as a miracle. Its GDP per capita increased from US$220 in 1980 to US$6000 in 2012 and the country became the world's second largest economy (Huang, Fang, Xu, &

L. Dai, *Politics and Governance in Water Pollution Prevention in China*, Politics and Development of Contemporary China,
https://doi.org/10.1007/978-3-030-02874-9_4

Qin, 2013). However, there is also consensus that the rapid economic growth exacted an enormous environmental cost (Wong, 2013). The Chinese Government has realized that a new, more sustainable development pathway is needed, which is associated with more sustainable development (Bakkes, Boer, Edelenbosch, & Vuuren, 2017). According to the Report delivered by Xi Jinping at the 19th National Congress of the Communist Party, "the entire [Communist] Party and the whole country have become more purposeful and active in pursuing green development, and there has been a clear shift away from the tendency to neglect ecological and environmental protection" (Xi, 2017).

This chapter looks at this recent shift which has led to a "greener" Chinese water law. It analyses in what way a sustainable, development-friendly approach to the utilization and management of water resources is reflected in China's domestic water law. It looks at water and environmental law in general, which are certainly highly correlated with water quality governance.

In the remainder of this chapter, two themes, sustainable development and the ecosystems approach, are introduced first (Sect. 2). Then these two themes are considered in connection with China's domestic water and environmental law and policy (Sect. 3). Section 4 discusses the recent institutional reform concerning environmental protection. This chapter concludes with Sect. 5.

2 Sustainable Development and Ecosystems Approach

2.1 Sustainable Development

The essence of sustainable development is that any development policy must equitably meet developmental and environmental needs of both present and future generations. This concept was developed in the framework of the United Nations. China, as a prominent member of the UN, played an important role in the concept's birth and subsequent evolution.

In 1972, the United Nations first adopted a common outlook and a set of 26 common principles for the preservation and enhancement of the human environment (United Nations, 1972). The concept of sustainable development was defined later in a report prepared by the World Commission on Environment and Development (Cassen, 1987).

The report called upon states to make their development policies more sustainable, and thus:

> To ensure that [their development policy] meets the needs of the present [generation] without compromising the ability of future generations to meet their own needs. (Cassen, 1987)

In 1992, the United Nations adopted a series of 27 principles of sustainable development ("Rio Declaration on Environment and Development", 1992). Some of the principles together constitute the essence of sustainable development:

(1) according to the principle of equity, "the right to development must be fulfilled so as to equitably meet developmental and environmental needs of present and future generations." (Principle 3)
(2) according to the principle of integration, "in order to achieve sustainable development, environmental protection shall constitute an integral part of the development process and cannot be considered in isolation from it." (Principle 4)
(3) according to the principle of public participation, "environmental issues are best handled with the participation of all concerned citizens." (Principle 10)

The first two principles, namely Principles 3 and 4, translate the essence of the concept of sustainable development into formulations with the potential to be solidified into legal principles. And Principle 10 is arguably one of the most important procedural principles related to sustainable development.

These general principles were translated into more concrete commitments by the Millennium Development Goals (MDG), derived from the Millennium Declaration, which included the goal of ensuring environmental sustainability (MDG7) (Rieu-Clarke, 2005; Spijkers & Honniball, 2015a, 2015b).

In recent years, China has taken the initiative in concluding a number of agreements encouraging the sustainable utilization of shared water courses. And at the domestic level, Chinese water law and policy have become more sustainable and environment-friendly.

2.2 *Ecosystems Approach*

It is difficult to give a definition of the ecosystems approach, since so much is still unclear about it. Generally speaking, an ecosystem is a biological unit or community of living components (plants and animals) and non-living components (soil, water, climate) interacting with each other (Lee, 2014). In the "Rio Declaration on Environment and Development" (1992), all states pledged to "cooperate in a spirit of global partnership to conserve, protect and restore the health and integrity of the Earth's ecosystem". The most authoritative description of an ecosystem in international law can be found in the "Convention on Biological Diversity" (1992). There, the term "ecosystem" is defined as:

> A dynamic complex of plant, animal and micro-organism communities and their non-living environment interacting as a functional unit. (Article 2)

To understand what freshwater ecosystems are, it is necessary to start at the Watercourses Convention. Article 20 provides that "watercourse States shall, individually and, where appropriate, jointly, protect and preserve the ecosystems of international watercourses" ("Convention on the Law of the Non-navigational Uses of International Watercourses", 1997). The International Law Commission, which is responsible for drafting Article 20, clearly linked this obligation to sustainable development, as the underlying aim was "to ensure [the ecosystems'] continued viability as life support systems, thus providing an essential basis for sustainable development" (International Law Commission, 1994).

According to the International Law Commission, an ecosystem is an "ecological unit consisting of living and non-living components that are interdependent and function as a community". Article 20 of ("Convention on the Law of the Non-navigational Uses of International Watercourses", 1997) established two different obligations for states, namely to: (1) protect and (2) preserve these ecological units. The obligation to protect "requires that watercourse States shield the ecosystems of international watercourses from harm or damage". The obligation to preserve "requires that these ecosystems be protected in such a way as to maintain them as much as possible in their natural state". It is as yet unclear whether Article 20 reflects existing customary law, binding also upon non-state parties such as China. This makes it even more interesting to analyse the practice of China.

3 Sustainable Development and Ecosystems Approach in China's Water Law and Policy

3.1 Sustainable Development in China's Domestic Water Law and Policy

"Sustainable development" has been extensively used as an umbrella term to link economic development, education and investment, and environmental protection in China's domestic policy. When it comes to natural resources management, it is stated in the Constitution that "the State ensures the rational use of natural resources" (Article 9). Signs of a more conscious environmental approach are shown by China's recent developmental blueprints and laws.

To implement the Sustainable Development Goals formulated by the United Nations, China formulated a National Plan on Implementation of the 2030 Agenda for Sustainable Development, which lays out guiding thoughts, a general roadmap and detailed plans for implementing the Sustainable Development Goals (The Central Government, 2016). In the Plan China commits itself to adhere to the basic national policy of resource conservation and environmental protection, follow a civilized development path, and pursue green development.

The recent 13th Five-Year Plan aims to promote a cleaner and greener economy, with strong commitments to environmental management and protection, ecological protection and security, and the development of green industries. Its sub-plan, the 13th FYP for Ecological and Environmental Protection (2016–2020) (The State Council, 2016), states that "development is of paramount importance but must be carried out in an appropriate way". It has laid out a series of environmental goals. In order to achieve these goals, the state commits itself to establish and put into practice a new philosophy of green development by upholding the fundamental state policy of conserving resources and protecting the environment when pursuing sustainable development. The goal is to build a resource-conserving, environmentally friendly society and bring about a new model of modernization whereby humankind develops in harmony with nature. Great importance has been attached to sustainable development in the past few years, a few examples showing this are mentioned below.

The state revised the Environmental Protection Law in 2014. The new law lays down the basic principle of coordinated development of economic construction, social development and environmental protection, and specifies that governments at all levels, all organizations and individuals have both the right and a duty to protect the environment. All of the three principles of sustainable development, set out in Sect. 2.1 above, are reflected in the new Environmental Protection Law. For example, Article 14 states that "when developing economic and technical policies, relevant departments of the State Council and people's governments ... shall take into full account their environmental impacts and solicit opinions from experts and relevant stakeholders" (the principles of equity and public participation). Article 13 states that "The governments ... shall include environmental protection work into their respective economic and social development planning" (the principle of integration).

The Water Law (2002) embodies the first two principles (namely equity and integration) as well. Like the Environmental Protection Law, it states that "the comprehensive watershed plan and comprehensive regional plan... shall be coordinated with the plan for national economic and social development, the overall plan for land utilization, the overall urban plan, and the plan for environment protection, and shall take account of the needs of the areas and industries" (Article 15, the reflection of the principle of equity); and "the development, utilization, preservation and protection of water resources, and the prevention and control of water disasters, shall be planned in a unified way on the basis of watersheds or regions" (Article 14, the reflection of the principle of integration). The Water Pollution Prevention and Control Law (2017) not only contains the first two principles, but also the principle of public participation. For example, the governments at or above the county level are required to "bring the protection of water environment into the national economic and social development planning" (Article 4, principles of equity and integration). It also prescribes that "any enterprise or public institution or any other producer or dealer that discharges poisonous and hazardous water pollutants ... shall release the information on poisonous and hazardous pollutants..." (Article 32, the reflection of the principle of public participation, see more in Chapter 3).

To prevent negative impact on the environment in the course of project construction, the state has enacted the Law on Environmental Impact Assessment to safeguard the country's sustainable development The Law states, as a purpose, that it is to promote "The carrying out of the strategy of sustainable development, prevent the unfavourable impacts of programmes and construction projects upon the environment after they are carried out, and promote the concerted development of the economy, society and environment" (Article 1).

A number of national strategic policies have been adopted as well. For example, the Action Plan for Water Pollution Prevention and Control requires governments to insist on the principle of participation by all people, promote that everybody is responsible for water saving and clean water, form a new mechanism for water pollution prevention and control which is to be led by governments, implemented by enterprises, driven by market and participated in by the public.

3.2 The Ecosystems Approach in China's Domestic Water Law and Policy

The 13th FYP is considered more comprehensive and stringent than the previous Plans; it represents a big step forward in the "war on pollution" (China Water Risk, 2016). It lays out the goal of achieving an overall improvement in the quality of the environment and ecosystems. The Central Government commits itself to step up ecosystem and environmental protection efforts by, for example, improving policies on function zones, promoting ecological and intensive resource use, and implementing the strictest possible environmental protection system, etc.

The Action Plan for Water Pollution Prevention and Control lays out specific objectives of water quality and with indices (Chapter 2). It aims to reach the goal that, by the middle of this century, the overall ecological environment quality will have been improved, and a virtuous cycle will have been achieved in the ecosystem.

The 2014 Environmental Protection Law, for the first time, sets an ecological red line for the strict protection of key ecological functional zones and areas of sensitive and fragile ecological environment. It stipulates that:

> The people's governments at various levels shall take measures to protect regions representing various types of natural ecological systems, regions with a natural distribution of rare and endangered wild animals and plants, regions where major sources of water are conserved, geological structures of major scientific and cultural value, famous regions where karst caves and fossil deposits are distributed, traces of glaciers, volcanoes and hot springs, traces of human history, and ancient and precious trees. Damage to the above shall be strictly forbidden. (Article 29)

Furthermore, the ecological protection compensation mechanism, which has been promoted by the State Council since 2005, also found its legal position in the new revised Environmental Law, which requires the state to "Provide guidance to the local governments of beneficiary areas and ecological protected areas on ecological compensation through consultation or market rules [and] increase fiscal transfer to areas of ecological protection" (Article 31, see more in Chapter 5).

There are many provisions in China's Water Law that reflect the ecosystems approach, although without explicitly using the word. For example, to protect water resources and improve the ecological environment, Article 9 of the Water Law requires China "To adopt effective measures to preserve vegetation, plant trees, grow grass, conserve water sources, prevent and control soil erosion and water pollution". Article 21 stipulates that "the development and utilization of water resources shall take into full consideration the ecological environment's need for water". These provisions clearly show that the Water Law is interested in the protection and preservation of more than just the watercourse itself: the entire ecosystem is included.

Compared with the 1998 version of the Water Law, the 2002 version added a whole chapter on water resource planning. As stated above, it requires the state to plan the development, utilization, preservation and protection of water resources, in a unified way, on the basis of watersheds or regions instead of administrative divisions. This provides the legal foundation of the Integrated Water Resource Management referred to earlier, which offers a lot of room for improving the water ecosystem.

The 2017 version of the Water Pollution Prevention and Control Law added "protecting the water ecosystem" as one of its legal purposes. It requires the relevant departments under the State Council and the local people's governments to "Make overall plans and take all factors into consideration, [to] maintain the reasonable water flow of rivers and the water level of lakes, reservoirs and groundwater at a reasonable place, [to] guarantee the basic water use for ecological purposes and [to] protect the ecological functions of waters when developing, utilizing, adjusting or transferring water resources" (Article 16).

And, in line with the Environmental Protection Law, Article 8 of the Water Pollution Prevention and Control Law states: "The state shall, in the mode of financial transfer payment or other means, establish a compensation mechanism for the ecological protection of the water environment in drinking water source reserve areas and upper reaches of rivers, lakes and reservoirs".

The 2015 Action Plan for Water Pollution Prevention and Control pays great attention to water-related ecosystem protection. For example, it states that "the capacity of water resources and the water environment will be given full consideration" (Article 2.6). It also lists the responsible sectors. This Action Plan sets out guidelines for future laws, as it states that "we will improve laws and regulations" and "reinforce law enforcement".

In earlier 2015, the Central Government published Opinions on Defining and Protecting Ecological Red Lines. The ecological "red line", also known as "ecological conservation red line", refers to strict space boundaries and environmental management limits that must be observed in ecological conservation services, environmental quality and safety controls, and natural resources utilization in order to ensure national and regional ecological safety, sustainable economic and social development, and people's health. It serves as both a "bottom line" and a "lifeline" in safeguarding national ecological safety. According to the Opinions, all regions need to decide on their "red lines" before 2020.

It is not only the laws which directly regulate water resources which can be linked to the ecosystems approach. Other laws, on similar topics, have also greatly contributed to the improvement of water ecosystems. For example, the Agriculture Law has a specific chapter on "the protection of agricultural resources and environment", in which the reasonable use of fertilizers and pesticides, the implementation of fishing norms, and the management of poultry and livestock cultivation, are all given due attention. These are the main contributors to the pollution of water and water ecosystems. The Land Administrative Law requires that the organizations which or individuals who develop unused land must do so "in accordance with the overall plans for land utilization, on condition that the ecological environment is protected and improved, and soil erosion and desertification are prevented" (Article 38). Besides their concern for ecosystem protection, both the Agricultural Law and the Land Administrative Law also identified sustainable development as one of the main purposes.

4 Institutional Reform Towards China's Green Path

The phrase "nine dragons rule the waters" is often used to refer to the fragmented roles and responsibilities for managing water resources in China (Dai, 2012). In 2015, the Central Committee of the Chinese Communist Party and the Chinese State Council promulgated an Integrated Reform Plan for Promoting Ecological Civilization. The plan listed a set of principles and objectives, and announced that, by 2020, China should complete the institutional set-up in line with ecological civilization. The Reform Plan spelled out a few key concepts of ecological civilization. It argued that "natural resources should be properly valued", and "holistically managed" (Ma & Liu, 2018). In the 13th FYP, the Central Government admits that the structural problems are becoming more evident and committed itself to making major progress in modernizing China's governance system and capacity for governance.

At the beginning of 2018, the most significant environmental governance reforms of this decade were taking place. A few new departments were formed, amongst which the Ministry of Ecological Environment and the Ministry of Natural Resources are particularly relevant to environmental protection and sustainable development.

The mission of the newly formed Ministry of Ecological Environment is, as stated on its official website, to improve environmental quality and build a beautiful China which enjoys blue sky, green land and clean water. It is responsible for the overall coordination, supervision and administration of major environmental issues. More specifically, it has adopted most of the responsibilities of the former Ministry of Environmental Protection; incorporated the function of underground water pollution regulation of the former Ministry of National Land and Resources; taken over the responsibility for watershed environmental protection from the Ministry of Water Resources; and the responsibility for agricultural pollution control from the former Ministry of Agriculture. The Ministry of Natural Resources has brought together the natural resources management responsibilities that used to belong to governmental departments in charge of forestry, agriculture, land resources, etc.

The reforms are broadly in line with the above-mentioned Reform Plan; the "holistic management" idea is reflected by both the Ministry of

Ecological Environment and the Ministry of Natural Resources. Experts are positive about the reform, they believe that "It will definitely lower the coordination cost among government agencies" (Ma & Liu, 2018). It can be expected that the implementation and enforcement of environmental laws and regulations will improve.

The government has also launched an international centre to advance research on sustainable development. Initiatives like these seem to illustrate that China is "rapidly moving from the chief polluter of the world to the chief provider of innovative, sustainable solutions for the world" (HSBC, 2017).

Not only is the government pursuing a more sustainable development approach, but also society and the public are playing their part. For instance, the Chinese Academy of Engineering, as a social organization, did comprehensive research on water issues, and published a collection of Strategic Research papers on the Sustainable Development of Water Resources in China in the twenty-first century, which has had a great deal of influence on water-related ideas and decisions of the government, academia and the public. The public is better informed than ever about China's water problems, from various sources, both governmental and from the media at large. According to an empirical study in 2014, 99% of the public believe that it is important to take urgent action to address and solve water issues, and 94% feel that they need to take more personal responsibility to help solve water issues in China. These are encouraging signs.

5 Conclusion

China has made enormous progress in recent years in making its water law and policy more sustainable development friendly. It has done so by developing a series of new development concepts, formulating and implementing national development strategies, adopting or revising national laws, and setting in place new institutional arrangements. The sustainable development approach and its principles, as well as the ecosystems approach, are reflected in various ways in many of the more recent legal and policy documents. It is a very promising start in terms of developing a "greener" Chinese water law framework.

References

Bakkes, J., de Boer, H. S., Edelenbosch, O. Y., & van Vuuren, D. P. (2017). *The Worldwide Context of China's Green Transition to 2050*. PBL Netherlands Environmental Assessment Agency.

Cassen, R. (1987). Our Common Future: Report of the World Commission on Environment and Development. *JSTOR*.

China Water Risk. (2016). *China's 13th Five-Year Plan for Ecological & Environmental Protection (2016–2020)*. Retrieved from http://chinawater-risk.org/notices/chinas-13th-five-year-plan-2016-2020/.

Convention on Biological Diversity, 2, Pub. L. No. Ch_XXVII_8. (1992).

Convention on the Law of the Non-navigational Uses of International Watercourses, No. 49 (A/51/49) Stat. (1997).

Dai, L. (2012). Recovering the Costs of Water Services in the People's Republic of China: Lessons from Article 9 of the European Union Water Framework Directive. *Utrecht Law Review, 8*(3), 102–118.

HSBC. (2017). *China's New Green Imperative*. Retrieved from https://www.bloomberg.com/news/sponsors/hsbc/chinas-new-green-imperative/?adv=7024&prx_t=0TYDArI8XAF0ANA.

Huang, Y., Fang, C., Xu, P., & Qin, G. (2013). New Normal of Chinese Development. In R. Garnaut, L. Song, & C. Fang (Eds.), *China: A New Model for Growth and Development* (pp. 35–54). Canberra: Australian National University.

International Law Commission. (1994). *Draft Articles on the Law of the Non-navigational Uses of International Watercourses and Commentaries Thereto and Resolution on Transboundary Confined Groundwater*. Retrieved from http://legal.un.org/ilc/texts/instruments/english/commentaries/8_3_1994.pdf.

Lee, J. (2014). *Preservation of Ecosystems of International Watercourses and the Integration of Relevant Rules*. Leiden: Brill Publishing.

Ma, T., & Liu, Q. (2018). *China Reshapes Ministries to Better Protect Environment*. Retrieved from https://www.chinadialogue.net/article/show/single/en/10502-China-reshapes-ministries-to-better-protect-environment.

Rieu-Clarke, A. (2005). *International Law and Sustainable Development*. IWA Publishing.

Rio Declaration on Environment and Development, United Nations Conference on Environment and Development, Pub. L. No. UN Doc. A/CONF.151/26/Rev.l (Vol. l) (1992).

Spijkers, O., & Honniball, A. (2015a). Developing Global Public Participation (1). *International Community Law Review, 17*(3), 222–250.

Spijkers, O., & Honniball, A. (2015b). Developing Global Public Participation (2). *International Community Law Review, 17*(3), 251–296.

The Central Government. (2016). *China's National Plan on Implementation of the 2030 Agenda for Sustainable Development*. Retrieved from Beijing, The Central Government.

The State Council. (2016). *13th Five-Year Plan for Ecological and Environmental Protection (2016–2020)*. Retrieved from http://www.stdaily.com/zhuanti01/guojia/2018-01/04/content_617866.shtml.

United Nations. (1972). *Declaration of the United Nations Conference on the Human Environment*. Retrieved from Stockholm.

Wong, E. (2013). https://www.nytimes.com/2013/03/30/world/asia/cost-of-environmental-degradation-in-china-is-growing.html.

Xi, J. (2017). Full Text of Xi Jinping's Report at 19th CPC National Congress.

CHAPTER 5

Exploring China's Approach to Implementing 'Eco-Compensation' Schemes—The Lake Tai Watershed as a Case Study

Abstract China is experimenting with various approaches to ensure the country's environmentally and socially sustainable development. Eco-compensation is one of the most influential initiatives. It aims to provide financial and economic incentives for environment management. This chapter examines China's approach to eco-compensation through experience in the Lake Tai watershed. Four typologies of eco-compensation schemes are identified and analysed, primarily through a legal lens. The author concludes that the main financial source for compensation is governmental payment and the form of compensation is limited to cash payments at current stage.

Keywords China · Eco-compensation · Lake Tai

1 Introduction

In their water governance framework, Van Rijswick et al. emphasize that empowerment with financial means is a crucial element for good water governance (van Rijswick, Edelenbos, Hellegers, Kok, & Kuks, 2014). The content that there are several ways to finance water management and several principles upon which to base the financing system.

China is experimenting with various approaches to ensure the country's environmentally and socially sustainable development. Eco-compensation is one of the most influential initiatives. It aims to provide

L. Dai, *Politics and Governance in Water Pollution Prevention in China*, Politics and Development of Contemporary China,
https://doi.org/10.1007/978-3-030-02874-9_5

financial and economic incentives for environmental management. This chapter selects the Lake Tai watershed as a case study and provides a legal analysis of the eco-compensation mechanism applied in the watershed and further provides an overview of how this mechanism has been applied in China. The Lake Tai watershed has been selected because it is: (1) one of the pilot eco-compensation schemes in China; (2) one of the most developed and polluted regions; and (3) one of the watersheds where a number and variety of eco-compensation schemes are being applied.

2 Eco-Compensation in China

The Chinese term "eco-compensation" (*sheng tai bu chang*) has often been used interchangeably with the international term "payment for ecosystem services" (PES), especially in comparative studies in this field examining China and global approaches (Zhang, Bennett, Kannan, & Jin, 2010). They share some similarities, but also have some differences. Internationally, PES is an important strategy in dealing with various challenges in environmental management. According to the UNECE, it refers to a contractual transaction between a buyer and a seller for an ecosystem service or for a land use/management practice which is likely to secure that service (UNECE, 2007). The Chinese notion of eco-compensation is broader, it covers an increasing range and number of national- and provincial-level innovations in environmental policy, and represents a potentially important new direction in China's evolving environmental policy framework (ADB, 2014). It encompasses both PES-like policies and also a wide range of other policies and programme types (Bennett, 2009). Generally, it creates not only incentives like PES, e.g. a reward or compensation for a right that is foregone in order to maintain a certain ecosystem service, but also disincentives, e.g. charges for the loss of or damage to ecosystems and natural resources, such as a pollution discharge fee (Li & Imura, 2007). Therefore, eco-compensation in China is defined in both narrow and broad terms. The narrow definition refers to the incentives and the broad definition covers not only incentives but also disincentives.

The eco-compensation mechanism was officially promoted by the State Council for the first time in 2005 and has developed very quickly since then. The State Council alone had published more than 146 regulations and policy documents concerning eco-compensation by 2017.

The National People's Congress Standing Committee, the legislative body in China, has revised more than 10 laws which touch upon the eco-compensation mechanism as well. The eco-compensation mechanism has been used in various fields such as water, forests, mineral resources, agriculture, and ecological function zones. Most of the mechanisms are directly or indirectly related to watersheds.

Although eco-compensation mechanisms are diverse in different watersheds, there are still some common features that can be distilled from state practice. The Lake Tai example provides an opportunity to learn valuable lessons for national water management regimes in the provision of water-related ecosystem services.

3 Eco-Compensation in the Lake Tai Watershed

Lake Tai is the third largest freshwater lake in China. The watershed occupies an area of some 3650 thousand hectares and extends across multiple jurisdictions: Jiangsu Province (52.6%), Zhejiang Province (32.8%), Shanghai Municipality (14%) and Anhui Province (0.6%). As one of the most developed regions in China, with only 0.4% of the land territory but 4.4% of the population, the Lake Tai watershed produced 10.3% of GDP; per capita GDP in this region was 2.4 times more than the national average in 2012 (Dai, 2014). The lake connects seven large cities across East China, including Shanghai and Hangzhou, which have a population of 23.8 million and 8.8 million, respectively (Dai, 2014).

In the 1960s and 1970s, Lake Tai enjoyed very good water quality (Class I–II), approximated to the water quality standards required in water source areas. In the late 1970s, the lake became seriously polluted because of the unprecedented economic growth and rapid urbanization in the regions. The entire lake has suffered from eutrophication since 1993, the most serious crisis arising in 2007 when many algal blooms, which were centimetres-thick, covered the entire lake and tap water turned yellow and was foul-smelling (Liang, 2012). Besides eutrophication, trans-regional water pollution problems are also persistent issues in this watershed since it extends across three provinces and one municipality. Both the central government and local government in the Lake Tai watershed have devoted themselves to improving water quality in the lake. Eco-compensation mechanisms, as one of the strategies, have been significantly developed across this region.

There are four types of eco-compensation used in the Lake Tai watershed: eco-compensation between governments; eco-compensation between governments and farmers; eco-compensation between governments and industry (Luo, 2011, pp. 82–89); and eco-compensation among industries. They are explored in more detail below (Dai, 2014). This chapter examines the practice of eco-compensation just in Jiangsu Province because the development of trans-provincial eco-compensation is still at an early stage.

3.1 *Eco-Compensation Between Governments*

Jiangsu Province selected four cities to test a bi-directional, inter-governmental eco-compensation mechanism between upstream areas and downstream areas from 2007 (Dai, 2014) and scaled it up to province-wide from 2014. To apply the bi-directional, eco-compensation mechanism, the provincial government establishes monitoring areas in boundary sections of rivers, lake entry sections, sea entry sections and trans-regional sections, and takes the water quality standards set by the provincial government or the central government (Chapter 2) as baselines, specifies the municipal and county governments as responsible bodies, and implements the compensation mechanism (Jiangsu Provincial General Office, 2013). According to the provincial government, the upstream areas need to compensate the downstream areas if the monitoring data in the trans-regional sections is worse than the water quality baseline, the rationale behind this is that the extra pollution caused by upstream areas results in extra expenditure on pollution control for downstream areas. Conversely, the downstream areas have to compensate the upstream areas if the monitoring data is better than the baseline.

There are 112 monitoring areas in Jiangsu Province, of which 30 are in the Lake Tai watershed. In these areas, three pollutants, namely Chemical Oxygen Demand (COD), Total Phosphorus (TP) and Ammonia Nitrogen are monitored. The provincial government sets a definitive unit price for each pollutant, and calculates the compensation amount according to the price and the monitoring data: the more the monitoring data deviates from the baseline, the higher the compensation. In order to motivate water quality protection, the provincial government applies a multiplier calculation approach. For example, if the monitoring water quality is 0.5 times worse than the designated baseline, the compensation price is RMB 250,000 per unit, but if the water quality is more than 0.5 times but less than 1 time worse than the baseline, the compensation price will be increased to RMB 500,000 per unit, and to RMB 1,000,000 per unit if the monitoring water quality is more than 1 time worse than the baseline. The provincial government also sets a reward scheme, e.g. the relevant government could get RMB 1,000,000 as a reward if it reaches the baseline for three consecutive years (Wu, 2014). Through these efforts, the water quality in the Lake Tai watershed has been greatly improved, progressing from Class V in 2007 to Class IV today, and for five consecutive years the Class V and inferior Class V water has been eliminated in the 15 rivers which flow into Lake Tai (Wang, 2017).

However, it should be noted that, under this type of eco-compensation, the compensation fund is not transferred directly between the upstream areas and the downstream areas, but goes to the provincial government's financial authority, which then re-distributes the fund to the relevant governments and requires that it is exclusively used for water environment treatment, water monitoring and water source protection. One salient feature of this type of eco-compensation is that the residents, communities and enterprises which are located in the upstream or the downstream areas are not taken into account despite the fact that their 'sacrifice' or 'contributions' are frequently mentioned in academic and public discussions (Wang, 2018). The provincial government is the dominant actor, from decision-making to exercising the mechanism in practice.

3.2 *Eco-Compensation Between Governments and Farmers*

As stated above, Lake Tai has suffered eutrophication outbreaks (blue algae) since the 1990s. The causes of eutrophication are Total Nitrogen (TN) and TP loads. If the TN and TP had not been included in the evaluation of the water quality, most of the surface water in Lake Tai would have reached Class III. However, when the TN and TP are taken into account, the water quality drops to inferior Class V, especially in Jiangsu Province. Sources of TN and TP in the lake are diverse. Intensive enclosed aquatic farms are one such source. In the East Lake Tai in Jiangsu Province, more than 90% of the surface water was occupied by enclosed aquatic farms (Dai, 2014). The Jiangsu provincial government, in order to achieve its water quality target for Lake Tai, has decided to remove all of those farms by the middle of 2019, which concerns around 3000 households of aquatic farmers, and covers around 450,000 Mu (30,000 ha) of surface water.

Jiangsu Province has drawn up and used regulations to address pollution problems caused by algal blooms since 2007. Removing the aquatic farms has always been one of the provincial strategies. However, problems arise in practice since the aquatic farmers are seriously affected by any such removal. From 2008 to 2018, the provincial government and the relevant municipal governments provided compensation subsidies through direct cash payments or aquatic farm resettlement for the farmers who suffered financial losses because of the clearance. However, the compensation procedure did not run smoothly, as the actual situation was very complicated. For example, since the municipal governments dominated the entire compensation process, with a lack of participation by interested parties, and with no third-party evaluations and assessments, it resulted in some poor decision-making, e.g. some of the newly resettled areas were not suitable for aquaculture (Dai, 2014).

Furthermore, the types of compensation were decided by the local governments as well. The local governments treated local farmers and non-local farmers differently. While the local farmers were permitted to select the compensation type, either direct cash compensation or resettlement, the non-local farmers had only one option, namely direct cash payments. This discriminatory approach led to protests by some non-local farmers. The situation of this unequal treatment was remedied in 2018 when a new policy was brought in. According to the new policy, the compensation of resettlement no longer existed because all of the aquatic farms have to be removed by 2019, there being no more space for resettlement. The situation of non-local, non-professional farmers has been improved as well. Under the new policy, both the local professional farmers and non-local, non-professional farmers can get not only direct cash compensation but also enjoy job skill training and be supported by social security. It is difficult to evaluate the effectiveness of this type of eco-compensation at present since the new policy has only just been brought in.

3.3 Eco-Compensation Between Governments and Industry

According to the Water Environment Comprehensive Management Plan for the Lake Tai Basin, there are some 2.10 million industries in the Comprehensive Treatment Region of the lake. Of this total, around 1.04

million are in Jiangsu Province and 1.06 million in Zhejiang Province, which not only contribute significantly to the economic development but also to the pollution in the Lake Tai watershed (Dai, 2014).

To control water pollution caused by the intensive polluting industries, the Jiangsu provincial government has implemented a discharge paid use scheme. Under this scheme, local governments allow their industries to discharge pollutants within a certain period after evaluating the receiving capacity of the surface water and setting a pollutants loading cap control for that water. Industries can buy discharge credits from the governments, with the preconditions of applying and paying for the discharge permits. This approach has been piloted on COD in Jiangsu Province since 2008; during the next two years, the purchased amounts of COD came to 49,700 tons per year, and the collected payments from discharge permits amounted to RMB175 million (US$28.2 million) (Li, 2010). The collected payments were used exclusively for environmental governance, the establishment of environmental monitoring and the construction and maintenance of the emission credits trading platform in the Lake Tai watershed within the Jiangsu Province jurisdiction. After seeing the success, the provincial government of Jiangsu extended the pollutants' scope from COD to a range of other pollutants, including Ammoniacal nitrogen (NH_3–N), Total Phosphorus (TP), total nitrogen (TN), sulphur dioxide (SO_2), oxides of nitrogen (NO_x) and Volatile Organic Compounds (VOCs).

Under this approach, the price of the emission rights is set by the provincial government. The amount of the pollution discharge is limited under the pollutants loading cap control system. It means that once the government has distributed all of the discharge credits for the pollutants, new applicants cannot purchase any credits from the government, but can only buy surplus credits from other dischargers via an emissions trading platform (see Sect. 3.4), or improve their own pollution prevention facilities to save credits themselves.

Nonetheless, it must be noted that this approach is different from the scheme of national pollution discharge fees. Under the scheme of discharge paid use, as described in the previous paragraph, governments set a pollutants loading cap, the amount of the pollution discharge is limited, which means that once the government has distributed all of the discharge credits for the pollutants, new applicants cannot purchase any credits from the government. Industries buy discharge credits guided by the principle of "the user pays", which reflects the dischargers' right to use natural resources. Under the scheme of national pollution discharge fees, however, dischargers pay fees when they discharge pollutants into the water. The difference between this scheme and the former scheme is that instead of governments setting a pollutants loading cap and allocating discharge credits, dischargers in the latter scheme report to and register with the local governments concerning the variety, quantity and density of discharged pollutants and wait for the governments' approval. Dischargers pay fees based on the principle of "the polluter pays"; it reflects the dischargers' liability for using the natural resources (Dai, 2014).

In comparing the two schemes, it seems that dischargers are more motivated under the scheme of discharge paid use than the scheme of national pollution discharge fees. This is because once they have saved discharge credits under the former scheme, they can keep them for the following year or sell them on the market (Sect. 3.4). Dischargers themselves, under this scheme, are the main pollution control bodies; governments only design and control the loading cap. Under the scheme of national pollution discharge fees, dischargers normally do not have enough motivation to reduce emissions if their discharges do not exceed the discharge standards approved by the governments. Governments are the main pollution control bodies, under which the cost of supervision is higher than in the discharge paid use scheme.

3.4 *Eco-Compensation Among Industries*

On the basis of the discharge paid use scheme, the emissions trading system has been initiated in a few pilot cities in the Lake Tai watershed since 2008, limited to COD emissions in the beginning. According to the provincial regulation, the industries could use, trade and pledge any surplus discharge credits after purchasing the emission rights from the governments; those surplus credits could be saved by applying cleaner production methods, improving pollutant discharge facilities, and technical reforms etc. In 2010, Jiangyin City was selected as a pilot city and out of 158 industrial dischargers in the city, 68 received extra discharge credits by emissions trading. The total turnover in the year was RMB 6.7 million (US$1.2 million) (Li, 2010).

In 2017, along with the expansion of the discharge paid use scheme, the scope of emission trading was extended to the above-mentioned pollutants as well, they are NH_3–N, TP, TN, SO_2, NO_x and VOCs. According to the Measures on Emission Rights Permission Rights Paid

Use and Trading, new industries could only get emission rights by emissions trading due to the cap control system, and this is also the case for old industries when they rebuild and extend their factories. Emissions trading occurs only in one administrative region, Jiangsu Province in this case, where the total maximum amount of pollutants is determined. Trading can only be initiated on a specified trading platform under the guidance and supervision of the environmental administration department, which is also responsible for examining the qualification of the transaction parties and formulating the trading rules. The revenue from this trading, as well as the money raised from the discharge paid use scheme, is used exclusively for environmental protection and treatment.

4 Discussion

The case study undertaken above reveals that four types of eco-compensation mechanisms have been deployed across the Lake Tai region, i.e. eco-compensation between governments, eco-compensation between governments and farmers, eco-compensation between governments and industry, and eco-compensation among industries (see Table 1).

In the first type of eco-compensation (between governments) the upstream government is required to compensate the losses of the downstream government when the water quality is worse than the defined quality in the monitoring areas. Conversely, the downstream government needs to pay the upstream government if the water quality is better than the expected level in trans-regional sections. Under this scheme, both broad and narrow eco-compensation are involved. The baseline of water quality at the trans-regional sections is determined by the provincial government and the central government, in other words, it is mandatory. Polluters that discharge pollutants exceeding the statutory level should bear legal liability. Therefore, the payment from the upstream government only is pollution charges: the purpose of this payment is for penalizing polluters, it can only be considered as eco-compensation in the broad sense. In the latter case, when the downstream government pays the upstream government because water quality is better than the statutory level, the downstream government actually pays for the additional ecosystem services: the payment is to reward the protectors and to motivate the service provider, the upstream government, in this case, to continually provide such service in the future. This approach works well in practice. The province has amassed more than RMB 1.3 billion

Table 1 Four types of eco-compensation mechanisms in Lake Tai region

Eco-compensation types	*Water-related services*	*Providers*	*Buyers*	*Payment*
Eco-compensation between governments	Provision of higher water quality than compulsory standards	Upstream city/ down-stream city	Downstream city/ upstream city	Financial transfer
Eco-compensation between governments and farmers	Improved water quality	Farmers	Governments	Cash, farm resettlement
Eco-compensation between governments and industry	Pollution control	Governments	Industries	Cash
Eco-compensation among industries	Pollution control	Industries	Industries	Cash

in its compensation fund between 2013 and 2016. The fund has been exclusively invested in water pollution prevention and control. The water quality in Jiangsu Province, including in Lake Tai, has been greatly improved.

However, under this scheme, the payment is not transacted between the direct polluters and beneficiaries, i.e. the residents or industries upstream and downstream, but only between the intermediaries, i.e. their governments. In the case of pollution charges from the upstream government, the fund principally should be collected mandatorily from the individual polluters, but it is not clear how this works in practice. The rewards paid by the downstream government is used exclusively in water pollution prevention and control. In other words, the rewards are not disbursed to the individual protectors. This leads to the problem that it may not be conducive to motivating the individual protectors to participate in water pollution control and thus can result in a funds shortage in the long run.

Under the eco-compensation scheme between governments and farmers, the governments compensate the farmers for changing their water use practices, which is aimed at improving water quality. The eco-compensation relationship seems to be comparatively clear—the ecosystem service buyers are the local governments, and the ecosystem service providers are the aquatic farmers (Table 1). However, since the buyers and providers are not on an equal footing, the transaction is not a voluntary one and problems have arisen in practice, such as the unequal treatment between local farmers and non-local farmers (see Sect. 3.2), and a low compensation rate set by the governments (Dai, 2014). Although it seems that the compensation prices have been increased in the new round of returning the aquatic farmers to the lake in 2018, the nature of the compensation is still not changed: i.e. the service buyer and provider are still unequal in status; the compensation scheme is still dominated by local governments; and farmers do not have any bargaining power due to the fact that they are in an unfavourable position. Therefore, this scheme is more of an administrative order than the form of eco-compensation which the government advocated, since the farmers involved in it have no other option but to accept the payment.

The third and fourth types, namely eco-compensation between governments and industry and among industries, have succeeded in making considerable contributions to various environmental protection funds. The discharge paid use scheme appears to work quite efficiently. Under this scheme, the governments play the role of ecosystem service providers for the purpose of maintaining a healthy water ecosystem and ensuring that the ecosystem can provide continuous eco-services, they set the pollutants loading cap for the region, monetize the pollutants and allocate the discharge credits. The selected dischargers are service buyers. However, under the emissions trading system, those selected dischargers become service providers, who save discharge credits and provide certain ecosystem services by improving their pollution prevention facilities or inputting some other efforts. Those who buy credits from other dischargers are service buyers (Dai, 2014).

However, despite their contributions, both the discharge paid use scheme and the emissions trading scheme have shortcomings. Firstly, in the case of Jiangsu Province, the provincial environment department, while making the rules governing the discharge paid use scheme, is also in practice the mediator and supervisor of the emissions trading scheme. With this administrative intervention, the market can barely play a role, which is not conducive to motivating the dischargers to participate in the emissions trading programme on the one hand, and on the other hand, there is also the risk of corruption since the environmental department is both the player and referee in this case, Furthermore, it is very difficult to evaluate the environmental benefits from the emissions trading scheme alone as it is generally applied together with many other policy instruments. According to research, tradable discharge permits are actually among the most challenging regulatory policies in terms of both their design and implementation (Kraemer, 2004, p. 11).

5 Conclusion

This chapter examines four types of eco-compensation schemes applied in Jiangsu Province to the Lake Tai watershed, and a number of observations can be made.

Eco-compensation schemes in the Lake Tai watershed are dominated primarily by governments through mainly government-sourced financial transfers (especially in the first three types). The main financial source for compensation is governmental payment and the form of compensation is limited to cash payments. Private parties are hardly involved in this at all. Although market-based eco-compensation has been examined within this region, it is still at a very early stage. A consequence of the single source of compensation is that it might lead to a funds shortage, which has happened with other, similar projects. This shortfall could, in turn, jeopardize the sustainability of the scheme. However, despite the existence of shortcomings, the current eco-compensation practice in Jiangsu Province reflects an appropriate policy innovation for the actual situation in China.

References

ADB. (2014). *Developing a Framework for Wetland Eco-Compensation Mechanisms in Jiangsu–Yancheng Coastal Wetlands.* Retrieved from http://www.povertyenvironment.net/sites/default/files/pep-files/Final%20Report%20-%20PRC%20-%20Jiangsu%20Yancheng%20Wetlands_0.pdf.

Bennett, M. (2009). *Markets for Ecoystem Services in China: An Exploration of China's "Eco-Compensation" and Other Market-Based Environmental Policies.* A Report from Phase I Work on an Inventory of Initiatives for Payments and Markets for Ecosystem Services in China, 86.

Dai, L. (2014). Exploring China's Approach to Implementing 'Eco-Compensation' Schemes: The Lake Tai Watershed as Case Study Considered Through a Legal Lens. *Water International, 39*(5), 755–773.

Jiangsu Provincial General Office. (2013). *Implementation Measures of Jiangsu Provincial Regional Eco-Compensation for Water Environment (Trial)*, 195 C.F.R. http://hbj.wuxi.gov.cn/doc/2014/01/21/387403.shtml.

Kraemer, R. A. (2004). *The Role of Tradable Permits in Water Pollution Control*, 11.

Li, L. (2010). *The Official Opening of Emission Trading Market in Tai Like Basin* (in Chinese).

Li, W., & Imura, H. (2007). *Eco-Compensation Mechanisms and Policies in China*. Beijing: Science Press.

Liang, G. (2012). Long Struggle for a Cleaner Lake Tai. *China Dialogue*. Retrieved from https://www.chinadialogue.net/article/4767-Long-struggle-for-a-cleaner-Lake-Tai-.

Luo, X. (2011). *Ecological Compensation Mechanism in Taihu Lake Basin: Based on Theory and Experience of Watershed Ecological Compensation In and Out of China*, 82–89 (in Chinese).

UNECE. (2007). *Recommendations on Payments for Ecosystem Services in Integrated Water Resources Management*, 51.

van Rijswick, H. F., Edelenbos, J., Hellegers, P., Kok, M., & Kuks, S. (2014). Ten Building Blocks for Sustainable Water Governance: An Integrated Method to Assess the Governance of Water. *Water International, 39*(5). https://doi.org/10.1080/02508060.2014.951828.

Wang, J. (2017, June). Water Quality and Water Ecology Has Been Improved in Tai Lake After Ten Years' Treatment. *China Daily*.

Wang, Y. R. (2018). *Characterising Water Eco-Compensation Mechanisms in China*. Sun Yat-sen University.

Wu, Z. (2014). *Jiangsu Province Promotes a Bi-directional Eco-Compensation in Watersheds*. Retrieved from https://www.thepaper.cn/newsDetail_forward_1269420.

Zhang, Q., Bennett, M. T., Kannan, K., & Jin, L. (2010). *Payments for Ecological Services and Eco-Compensation: Practices and Innovations in the People's Republic of China*. Asian Development Bank.

CHAPTER 6

Implementing the Water Goals—The River Chief Mechanism in China

Abstract Water governance assessments often focus mainly on the beginning of the policy process. There is ample attention for formulating goals, rules and standards. However, good water governance should pay attention to the whole policy process from goal setting to the actual achievement of goals. More attention is needed for the implementation of regulations. Chinese Central Government has launched a creative approach to implement its water goals—the river chief mechanism, by which the local government heads are appointed as river chiefs to clean up and protect water resources in their jurisdictions. This chapter intends to analyze this mechanism through discussing its origins and evolution, the rationale behind the mechanism and the pros and cons of this mechanism.

Keywords China · Water governance · Water quality · River chief

1 Introduction

Water management and water governance assessments often focus mainly on the beginning of the policy process. There is ample attention paid to public participation, formulating goals, rules and standards, and the process of decision-making. However, good water management and governance should pay attention to the whole policy process, from goal setting to the actual achievement of goals (van Rijswick, Edelenbos, Hellegers,

L. Dai, *Politics and Governance in Water Pollution Prevention in China*, Politics and Development of Contemporary China,
https://doi.org/10.1007/978-3-030-02874-9_6

Kok, & Kuks, 2014). Therefore, more attention needs to be directed to the implementation of regulations.

The Chinese Central Government has launched a creative approach for implementing its water goals—the river chief (*He-Zhang*) mechanism, by which the local government heads are appointed as river chiefs and are responsible for the management and protection of watercourses, as well as preventing pollution and restoring the water ecology. This chapter will introduce the mechanism (Sect. 2), analyse the rationale behind the mechanism and its evaluation tools (Sect. 3) and discuss the strengths and weaknesses of this mechanism (Sect. 4). Section 5 provides a conclusion.

2 The River Chief Mechanism

The river chief mechanism originated from Wuxi City in Jiangsu Province. In 2007, a devastating blue algae bloom in Lake Tai seriously affected Wuxi City. To tackle water pollution caused by the blue algae, the municipal government adopted a series of measures and the river chief mechanism came into being. Under this mechanism, the heads of the Wuxi Communist Party and government officials were appointed as "chiefs" of the rivers and tasked with water quality protection.

Under the Wuxi municipal river chief mechanism, officials who achieve the assigned annual targets of water quality control could get rewards. Those whose rivers show signs of pollution are punished with fines or loss of promotion opportunities. The Communist Party of China (CPC) Wuxi Committee incorporated the performance of the chiefs into a veto system, under which a failure to attain the assigned targets will unilaterally lead to the disqualification of the officials (Dai, 2015). As stated by the CPC Wuxi Committee:

> "For those who poorly govern environmental protection, those who cannot achieve the established objectives [...]", the Organizational Department (of the CPC), after an investigation, will "veto the relevant leadership when they participate in a city-level competition for an effective leading group, or veto the relevant responsible officials when they participate in the competition for advanced or excellent individuals, or veto the promotion of those directly responsible officials [...]". (Tai Hu Net, 2016)

Under the veto system, the river chiefs experience significant career pressure, which persuades them to devote themselves to protecting the rivers for which they are responsible. As a result, the water quality in Wuxi City was greatly improved within a short period. The river chief mechanism therefore rapidly became a popular tool for tackling water pollution, first across the province and then across the country. The Jiangsu provincial government began to promote and improve the mechanism from 2008, and decided to expand it to five administrative levels in 2017, namely province, city, county (district), town (community) and village levels. By the end of 2017, the mechanism in Jiangsu Province alone has covered over 100,000 rivers in villages and more than 20,000 rivers in counties (Dylan, 2017). Meanwhile, some other provinces have also achieved satisfactory results from trialing the application of the river chief mechanism (Xu, 2017).

At the national level, the Ministry of Water Resources has been encouraging the development of the river chief mechanism nationwide since 2010. In 2016, the CPC Central Committee and the State Council jointly published the official Guidance for Full Implementation of the River Chief Mechanism, stipulating that the mechanism should be fully established throughout the country by the end of 2018. By then, China will have established river chief mechanisms at the provincial, municipal, county and township levels, with support at the highest (Presidential) level. The heads of local government at the various levels are appointed as river chiefs and will be held accountable not only for water pollution control, but also resource protection and ecological restoration in their jurisdictions. Their performance as river chiefs will be evaluated as part of their political performance (*China Daily*, 2016).

Up to the end of 2017, there have been more than 900,000 river chiefs appointed at each level of government nationwide (Gu, 2017). Assessment measures for river chiefs have been set up as well and these are discussed in the following section.

3 Rationale Behind and Evaluation Tools of the River Chief Mechanism

3.1 Rationale Behind the River Chief Mechanism

Structurally, China's bureaucratic ranking system combines with the functional division of authority among various government departments to produce a situation in which it is often necessary to achieve agreement among an array of bodies (Lieberthal & Lampton, 1992). However, departments often remain isolated from one another, no single one has authority over the others. Furthermore, tasks are often ambiguously divided among different departments; it is often the case that those responsible for the actions taken cannot be traced. In the case of water management, there are many governmental departments involved. From an administrative perspective, there are national ministries like the Ministry of Water Resources and the Ministry of Ecology and Environment, provincial departments and urban bureaus. From a natural perspective, there are river basin commissions, some of which can span provinces. From a functional perspective, there are different government departments in charge of specific issues such as water pollution, water resource, floods, water fees, etc. (Xu, 2017). Empirical findings have provided evidence that this fragmentation has a detrimental effect on the institutional capacity of the governance system to respond to water quality degradation, both in respect to pollution accidents and to the slower processes of deterioration caused by human activity (Silveira, 2014).

In order to overcome these institutional hurdles, governments at all levels are making continuous efforts by, for example, improving legislation and establishing legal systems. For example, the Water Law, which was amended in 2002, clearly stipulated that the main function of water resources management is to combine river basin management with administrative regional management so that together they can provide an integrated approach to water resources management (Global Water Partnership, 2015). However, there is a mixed picture in terms of the outcomes: some evidence shows that significant improvements have been achieved, while others question the applicability (Yu, 2014). For example, Biswas (2008) argues that western-based approaches, such as Integrated Water Resource Management, are not workable for large countries such as China.

In terms of law implementation, China still faces severe challenges. Environmental regulations were worded "vaguely", leaving "considerable scope for interpretation", and "contradictions" were observed in the assignment of implementation authorities among different branches and levels of government (Yee, Tang, & Lo, 2014). Furthermore, a fully mature legal system has not yet been established at this stage; executives still base their environmental practices largely on perceptions about regulators rather than concerns for legal issues (Yee et al., 2014). To solve this problem, China has established a target-oriented responsibility system, i.e. from the centre down to township-level governments, higher-level authorities set up targets and cascade them down to lower levels of governments, forming a target pyramid. The work performance of chief executives in target fulfilment was internally ranked on a collective as well as an individual base, and bonuses were paid or punishments imposed accordingly (Burns & Zhou, 2010). This system has recently been enshrined in law and in the regulations, through which the higher-level governments could enable lower governments to regulate the behaviour of their employees and make sure that the government departments fulfil their responsibilities.

In the field of water management and protection, the government's chief officers at and above the county level assume overall responsibility for managing and protecting water resources. The State Council checks and inspects each province, autonomous region and municipality using appropriate indicators (Global Water Partnership, 2015), combined with incentive approaches, to ensure the country's water targets are fulfilled.

The river chief mechanism is designed to implement the target-oriented responsibility system. Under the river chief mechanism, the principal leaders of party committee and government are usually assigned to take charge of and supervise the rivers and lakes in their jurisdictions. For example, at the provincial level, the Party Committee Secretary and Governor are usually appointed as principal chiefs in charge of the rivers and lakes within the province. It should be noted that the Party in China has formal political authority over the government, the Provincial Party Committee Secretary therefore has the highest rank in a province, and the Governor is next in rank. Both the Party Committee Secretary and Governor have a strong influence over the divisions of responsibility in their governmental departments and the power to mobilize different departments to coordinate and build consensus in specific areas in a short period. Furthermore, since their performance is also evaluated by the central government, which usually induces them to follow central policy and decisions by using promotion to central department or higher positions as an incentive, they are also motivated to perform well.

Therefore, the river chief mechanism has been able to achieve a level of success in a short period because, in summary, the mechanism enables both horizontal and vertical integration and, to a large extent, solves the problem of institutional fragmentation. Horizontally, it brings together representatives from multiple governmental departments to coordinate on water problems under the guidance of the principal river chiefs; in this way the sectorial strength concerning pollution control and resource management can be combined in a short period of time. Vertically, governors, mayors, county heads, district heads, township heads and village heads all serve as river chiefs at all levels, which can help to improve the information flow about river and water quality between upper and lower levels of government.

3.2 *Evaluation Tools of the River Chief Mechanism*

According to the Guidance for Full Implementation of the River Chief Mechanism, the performance evaluation standards of the river chiefs in different regions are drafted differently in accordance with their specific problems. The central government suggests that local governments should apply the natural resource-based balance sheet and auditing system when assessing the performance of the chiefs.

The natural resource-based balance sheet and auditing system is an innovative approach being promoted by the central government; it reviews, via an auditing system, the fulfilment of environmental duties by cadres before they leave their posts. When evaluating river chiefs, water quality deterioration and water pollution incidents which had occurred during their time in office would both be reviewed. The purpose is to establish the liability for damage to the ecological environment by means of a lifelong accountability system, which has been used in some pilot cities since 2015 and has been upscaled nationwide since 2017. However, the preparation of the natural resource-based balance sheet is very complex and challenging in practice. For example, it is difficult to calculate the value of the services provided by the natural resources being audited on the one hand and, on the other hand, dividing the liability between the predecessors and the officials currently in office is also difficult because environmental effects often appear only after a long period of time, meaning that the effects may not become apparent during a five-year tenure of leadership. Furthermore, the auditing system per se still has some deficiencies (Guo, 2018).

In practice, the veto power is widely used to evaluate the performance of the river chiefs. The veto power was originally applied to birth control in Southern China in 1984 and was then updated as a national tool to ensure social stability after 1991. Energy saving and emission reduction came within the remit of the veto power in 2007 (Dai, 2015). Targets with a veto power are used for key policies at central levels and sometimes also used for key policies at local levels (Edin, 2003). These targets are the most important for cadres and explicitly linked to personnel decisions. For example, if an official fails to meet these targets, he or she would not be evaluated for annual bonuses or promotion, or could even face a salary cut or dismissal, which puts great pressure on local officials, especially for cadres in the leadership corps who are directly accountable for the failure of veto targets on behalf of their organizations (Burns & Zhou, 2010). Poor performance concerning veto targets cannot be remedied by good performance on other targets (Wang, 2013).

To ensure compliance, local governments also dispatch inspection teams to check up on officials after setting the targets. For example, Huanggang Municipal Government (of Hubei Province) signed responsibility contracts with its county governments on environmental protection targets in 2009. According to these contracts, the Municipal Party Committee Secretary serves as the principal river chief and is in charge of the rivers in his region, and his political partner, the Mayor, acts as the principal lake chief in charge of the lakes in the city; the heads of counties and towns act as either river chiefs or lake chiefs in their corresponding jurisdictions. The performance of target fulfilment by those chiefs is investigated by the municipality on a monthly, quarterly, half-yearly and yearly basis. If the monthly targets are not fulfilled, the local chiefs will be warned by the municipality. If the local chiefs fail to achieve the quarterly targets, they will be given a public reprimand. If the half-yearly targets are not achieved, the failure will be recorded in the personal profiles of the local chiefs, and if the yearly target is not met, the veto power will be applied (Zhou, 2009). Correspondingly, the performance of the principal river and lake chiefs is investigated by the higher-level governments; thus Hubei Province, by means of a similar system, and chiefs at the provincial level are investigated by the central government. All of the investigation results, both positive and negative, affect the chiefs' careers to some extent. The career pressure resulting from this investigation process, especially the veto power, provides a key incentive that drives the river chiefs to work hard to achieve their assigned tasks.

4 Strengths and Weaknesses of the River Chief Mechanism

The river chief mechanism, as an adaptive tool to deal with the urgent pressures of water problems, has its strengths, but also its weaknesses.

4.1 *Strengths of the River Chief Mechanism*

Firstly, the river chief mechanism, as stated above, largely overcomes institutional hurdles and improves administrative efficiency, especially for the management of large rivers that span regions, in which fragmentation has been a long-standing problem. Under the river chief mechanism, principal river chiefs, while being responsible for leadership, also have the responsibility for organizing the river chiefs conference and coordinating with their upstream, downstream, left and right bank authorities to find solutions to outstanding problems, thereby ensuring joint protection and management. Information sharing between different administrative regions could also be strengthened by a joint conference, by which the exchange of data and the implementation of progress on the shared rivers are facilitated. Furthermore, the river chief mechanism, by making the official leaders of the party and governments the first responsible actors, does not change the function of different departments. From this point of view, it is very cost efficient since there is no need to create new institutions or positions.

Secondly, the river chief mechanism promotes site-specific policies. In a country like China, water pollution sources vary on a regional basis, for example, industrial and agricultural pollution, two main pollution sources, are unevenly distributed. National or provincial regulations are not always suitable for local conditions. The river chiefs, however, can compensate for this problem. By diagnosing the specific causes of water pollution in their rivers, the chiefs could relate the remedy to the specific situation, so that pollution problems are dealt with on a river-based perspective; the remedy therefore tends to be more practical (Dai, 2015).

Thirdly, the river chief mechanism promotes law implementation. Previously, environmental achievements were not taken as an indicator for assessing officials' performance (Dai, 2015), which greatly reduced the motivation of lower governments to conduct environmental protection. The enthusiastic pursuit of economic growth led governments to ignore the legal norms on a wide scale. For example, in a random check of 126 industrial parks in 11 provinces (autonomous regions) in 2007, 87.3% of these parks were shown to have been approved by governments illegally or ultra vires, or by lowering the standards of the environmental impact assessment (Dai, 2015). The implementation of the river chief mechanism, taking water quality as a compulsory indicator in the assessment of the chiefs, should improve this situation, since the implementation of existing laws and regulations is usually a shortcut to achieving the water goals.

4.2 Weaknesses of the River Chief Mechanism

Although the river chief mechanism could promote law implementation on the one hand, on the other hand it could also jeopardize the principle of the rule of law. In practice, the mechanism is largely controlled by individuals, success and failure to a great extent depends on who the river chiefs are. Officials work hard to fulfil the assigned tasks, very often because of the career pressures, such as salary and career rewards or sanctions. However, the rewards cannot always steer officials to accomplish the targets because, for example, every river chief has his own understanding of what career achievement means. It is not the case that every government official has the same enthusiasm concerning the pursuit of his career, especially when he may be close to the age limit for promotion. As one local official once said, he was "too old for promotion, too young to stop working, but just right for mahjong and drinking" (Smith, 2013). Therefore, under the river chief mechanism, it is possible to have tight control on water quality at one time or in one place and to have a less rigid policy in others due to the individual character and situation of the chiefs (Dai, 2015).

Furthermore, the core operational tool, the veto power of the river chief mechanism, is not only applied in the river chief mechanism, but also in other fields; for example, GDP achievements, safe production control and food safety. According to a news report, in Jiangxi Province alone there were 62 different items under the veto system applied to its local governments up to 2011 (Pan, 2013). Local cadres have a wide range of responsibilities, water pollution control being only one of them. Officials are often overburdened with pressure, especially when each of the responsibilities plays a decisive role in an official's promotion, reward or sanction. This pressure tends to lead to favouritism and manipulation which are unavoidable under certain circumstances especially when competition for promotion is fierce. Many strategies have been adopted by lower governments in order to cope with the overwhelming pressure, for example, selective reports, selective implementation or the manipulation of data, etc. (Dai, 2015).

Thirdly, although the river chief mechanism tends to be a long-term strategy, it is not immune from the allure of campaign-style enforcement. Unprecedented efforts have been made in water quality control ever since the implementation of the river chief mechanism, the problem of controlling water quality at any cost has shown up in many places. For example, in some counties and cities, although their annual fiscal revenue is less than 1.5 billion RMB, the annual budget for water pollution control has exceeded this amount. This inevitably led to a significant increase in the debt ratio of local governments (Shen, 2018).

Last but not least, public participation is not well developed under the river chief mechanism and although progress has been made in recent years, it is nevertheless still at the stage of limited participation (Chapter 3). This is mainly because the target-oriented responsibility system was designed primarily for governmental internal hierarchical control rather than outward public accountability. Performance measurement was primarily an internal exercise within the government system, initiated, executed and used by the government (Burns & Zhou, 2010). At the beginning of 2014, the Ministry of Water Resources, jointly with 10 other ministries, issued an Implementation Plan on the Most Stringent Water Management Assessment, in which the process, contents and scoring methods of how to assess local governments are specified (Dai, 2015). However, It is not yet clear how this Plan can be harmonized with the veto system. Since the veto system is the Party's personnel management tool, internal reform of the Party is needed for the Plan to work in practice.

5 Conclusion

This chapter introduces the river chief mechanism, analyses the target-oriented responsibility system and the veto power which together make the mechanism operational, and discusses the strengths and weaknesses of the mechanism.

In general, although the river chief mechanism is largely ruled by individuals and seems to conflict with the country's objective of building a society ruled by law, it is still an efficient tool for implementing water goals designated by the central level and for curbing emerging environmental pressures during this transitional period. The mechanism has been developed against the background that the Party runs the state through a political system with a tight, top-down approach, in which the law is often less important than Party policy. Therefore, while strengthening the legal framework on the one hand, on the other hand there is certainly room for improving and a need to improve this mechanism from the Party's side, for example, by increasing transparency and public participation, the implementation of which, however, requires a change in Party mentality.

Furthermore, from both domestic and international river governance experience, improving environmental quality, such as water pollution treatment and water ecology restoration, which now are the responsibility of the river chiefs, often takes decades. This suggests that to make the river chief mechanism longlasting in its effect, the country should also integrate the understanding of water protection needs into the public education system.

References

Biswas, A. K. (2008). Integrated Water Resources Management: Is It Working? *International Journal of Water Resources Development, 24*(1), 5–22. https://doi.org/10.1080/07900620701871718.

Burns, J. P., & Zhou, Z. (2010). Performance Management in the Government of the People's Republic of China: Accountability and Control in the Implementation of Public Policy. *OECD Journal on Budgeting, 2*, 1–28.

China Daily. (2016). River Chief System. Retrieved from http://www.chinadaily.com.cn/opinion/2016-12/13/content_27649825.htm.

Dai, L. (2015). A New Perspective on Water Governance in China: Captain of the River. *Water International, 40*(1), 87–99. https://doi.org/10.1080/02508060.2014.986702.

Dylan. (2017). *Jiangsu to Implement 'River Chief' System*. Retrieved from http://english.jschina.com.cn/23261/201703/t20170329_3866484.shtml.

Edin, M. (2003). Remaking the Communist Party-State: The Cadre Responsibility System at the Local Level in China. *China: An International Journal, 1*(1), 1–15.

Global Water Partnership. (2015). *China's Water Resources Management Challenge: The 'Three Red Lines'*. Sweden: Elanders.

Gu, L. (2017). *China Pushes 'River Chief' System to Combat Water Pollution*. Retrieved from http://www.ecns.cn/2017/12-20/285138.shtml.

Guo, X. (2018). Major Problems and Causes Analysis on Natural Resource-Based Balance Sheet and Auditing System of Leading Cadres in F Province. *China Chief Financial Officer, 4*, 122–123.

Lieberthal, K. G., & Lampton, D. M. (1992). *Bureaucracy, Politics, and Decision Making in Post-Mao China*. Berkeley: University of California Press.

Pan, J. (2013). *Veto Should Not Become a Norm of Cadre Evaluation*. Retrieved from http://epaper.gmw.cn/gmrb/html/2013-11/07/nw.D110000gmrb_20131107_2-02.htm.

Shen, M. (2018). *Analysis on the River Chief System from the View of Institutional Economics*, 134–139.

Silveira, A. (2014). China's Political System, Economic Reform, and the Governance of Water Quality in Pearl River. In D. E. Garrick, D. Connell, &

J. Pittock (Eds.), *Federal Rivers: Managing Water in Multi-layered Management Systems* (384pp). Cheltenham: Edward Elgar.

Smith, G. K. (2013). Measurement, Promotions and Patterns of Behavior in Chinese Local Government. *Journal of Peasant Studies, 40*(6), 1027–1050.

Tai Hu Net. (2016). *Modern River Chief Mechanism Expended to Nationwide from Here*. Retrieved from http://www.tba.gov.cn/tba/content/TBA/xwzx/slyw/0000000000011409.html.

van Rijswick, H. F., Edelenbos, J., Hellegers, P., Kok, M., & Kuks, S. (2014). Ten Building Blocks for Sustainable Water Governance: An Integrated Method to Assess the Governance of Water. *Water International, 39*(5). https://doi.org/10.1080/02508060.2014.951828.

Wang, Y. (2013). *Public Participation and Legal Reform Key to Creating Ecological Civilisation*. Retrieved from https://www.chinadialogue.net/article/show/single/en/6583-Public-participation-and-legal-reform-key-to-creating-ecological-civilisation.

Xu, Y. (2017). *China's River Chiefs: Who Are They?* Retrieved from http://chinawaterrisk.org/resources/analysis-reviews/chinas-river-chiefs-who-are-they/.

Yee, W.-H., Tang, S.-Y., & Lo, C. W.-H. (2014). Regulatory Compliance When the Rule of Law Is Weak: Evidence from China's Environmental Reform. *Journal of Public Administration Research and Theory, 26*(1), 95–112.

Yu, H. H. (2014). Community-Based Water Governance Under Integrated Water Resources Management Reform in Contemporary Rural China. *Environmental Management and Sustainable Development, 3*(2). https://doi.org/10.5296/emsd.v3i2.5656.

Zhou, C. (2009). *Tackling the Pollution via River Chief Mechanism, Solutions of Environmental Contract Responsibility System*. Retrieved from http://finance.ifeng.com/a/20090409/521685_0.shtml.

CHAPTER 7

Summary and Conclusion

Abstract China has made considerable policy progress on building the regulatory framework of water quality management during the last decade. Longer-term pollution strategy has been set out at the central level, considerable attention has been given to green development. There are many promising activities carried out by the Central Government during last decade. For example, declaring war against water pollution, revising water laws, promoting public-interest litigation, developing market-based approach, carrying out institutional reform and strengthening the implementation etc. However, gaps between expectations and performance still exist, more attention should be given to implementation in the future.

Keywords Water management · China · Water pollution · Public-interest litigation

1 Introduction

This book set out to explore the governance in water pollution prevention and control in China. The preceding chapters offered a reflective analysis from different perspectives according to a three-step diagnostic method and the "ten building block" assessment approach. Through this approach, China's water quality status, water policy discourse, public

L. Dai, *Politics and Governance in Water Pollution Prevention in China*, Politics and Development of Contemporary China,
https://doi.org/10.1007/978-3-030-02874-9_7

participation in water quality management, green transitions considering water laws, the interaction between state and non-state actors in the water sector and the implementation of water goals are discussed.

2 Reflections on and a Synthesis of the Findings

2.1 The Content of Water Governance in China

The content of water governance in this book encompasses water quality knowledge (Chapter 2), water policy discourse (Chapter 2) and public participation (Chapter 3).

The status of water quality is demonstrated by yearly environmental bulletins in China, which form a basis for water quality management. Currently, China's water pollution problems are still serious although there has been an improvement compared with a decade ago. The central government has recently intensified efforts to formulate action plans and programmes to control and remedy water pollution, in which detailed objectives and standards are laid down. Binding targets have been set for particular pollutants and positive results have been achieved (Chapters 2 and 6). However, too much focus on the binding targets has led to other important pollutants, which do not have binding targets, being neglected. As a consequence, the decline in pollutants without binding reduction targets has been much lower in comparison with the decline in the four pollutants with binding targets, namely, SO_2, COD, Ammonia Nitrogen and Nitrogen Oxides.

The Chinese Government has realized that public participation is a potential tool to enhance effective environmental governance and it has actively promoted environmental public participation in its recent policy and legal documents, i.e. the Five-Year Plans (FYP), national laws and regulations, which serve as a mandate for the public to participate. However, obstacles remain to substantial participation in practice, especially under the river chief mechanism (Chapter 6), because, on the one hand, details of the conditions and procedures for public participation have not yet been clearly stipulated and, on the other hand, there is not yet a mature platform for the public to participate in the Party's internal decisions, which largely influence water governance in China. The new revised Environmental Law is an important stride forward in terms of promoting public participation. It provides qualified NGOs with legal status for public interest litigation, which is a promising start although there are still many challenges in practice. It is expected that there would be new legislation or policies which could lend further support to NGOs.

2.2 The Organization of Water Governance in China

Under the organization of water governance, two important elements, i.e. the transition towards a "greener" Chinese water law (Chapter 4) and the eco-compensation mechanism (Chapter 5), are discussed.

The Chinese central government has made increasingly intensified efforts towards sustainable development in recent years. The government has signaled clearly by recent FYP and actions that it lays emphasis on green development. Its guiding policy on water management and the philosophy of sustainable development have been codified by newly revised laws and regulations. The central government has committed itself to stepping up ecosystem and environmental protection efforts by, for example, implementing the strictest possible environmental protection system. Institutional reform is taking shape as well. Two new ministries, the Ministry of Ecological Environment and the Ministry of Natural Resources, have been recently formed for the purpose of better serving the country's environmental protection. This new round of institutional reform is significant in China: the idea of holistic management reflected therein, if implemented, will end the longstanding problem of institutional fragmentation and put China on the road to sustainable development in the next decade.

Eco-compensation is one of the most influential initiatives with which China is experimenting. In Chapter 5, Lake Tai was selected as a case study to examine the implications of the eco-compensation mechanism at the local level. It found that four types of eco-compensation are being developed and applied across this region, namely eco-compensation between governments, eco-compensation between governments and farmers, eco-compensation between governments and industry and eco-compensation among industries. The common feature in each type is that the government plays a dominant role, the main financial source for compensation being government payments. Governments are the main actors in formulating and implementing the scheme. Although market-based eco-compensation has been examined within the Lake Tai region, it is still at a very early stage.

2.3 *The Implementation of Water Governance in China*

Chapter 6 discusses the implementation of water goals in China. It analyses a popular policy instrument—the river chief mechanism—to show how China applies its target-oriented responsibility system as an institutional tool to address water quality challenges and to fulfil the tasks assigned by higher level governments. From the moment that water quality was set to be enhanced by means of a national binding target, the river chief mechanism rapidly proved to be an adaptive and effective tool to fulfil this target. Through applying the river chief mechanism, administrative efficiency has been greatly improved in the short term, and the pressure caused by fragmented institutions has been temporarily mitigated. However, the mechanism also has its shortcomings, for example, the mechanism's assessment measure relies largely on the veto power, although other measures are being developed, such as an auditing system and a lifelong accountability system. This has caused some problems in practice, such as the unexpected countermeasures, the campaign-style enforcement, etc.

3 Conclusion

Since the 11th FYP, China has made considerable policy progress on building the regulatory framework of water quality management. Longer term pollution strategy has been set out at the central level and considerable attention has been given to green development. There have been many promising activities carried out by the central government during the last decade. For example, declaring war against water pollution, revising water laws, promoting public interest litigation, developing a market-based approach, carrying out institutional reform and strengthening implementation. However, gaps between expectations and performance still exist and difficulties still remain in the actual implementation. There is an old Chinese saying: the law cannot work on its own, which means that drafting a law, issuing a policy or establishing an organization does not guarantee that the law will be effective in practice: the law must be applied and enforced. Therefore, more attention should be given to implementation in the future.

Index

L. Dai, *Politics and Governance in Water Pollution Prevention in China*, Politics and Development of Contemporary China, https://doi.org/10.1007/978-3-030-02874-9

The manufacturer's authorised representative in the EU is Springer Nature Customer Service Centre GmbH, Europaplatz 3, 69115 Heidelberg, Germany. If you have any concerns regarding our products, please contact ProductSafety@springernature.com

Printed and bound by CPI Group (UK) Ltd, Croydon, CR0 4YY

30/09/2026

02225466-0001